War Machine

THE RATIONALISATION OF SLAUGHTER IN THE MODERN AGE

DANIEL PICK

YALE UNIVERSITY PRESS   NEW HAVEN & LONDON · 1993

To Isobel and to Eric

Set in Sabon by Best-set Typesetters Ltd, Hong Kong
Printed and bound in Great Britain by The Bath Press, Avon

Contents

List of Illustrations

Acknowledgements

I would like to thank Roy Porter, Tony Tanner, Jay Winter, Gareth Stedman Jones, Jacqueline Rose, Carla Rachman, Simon Schaffer, Irma Brenman, Jennifer Davis, Margaret Hanbury and Robert Baldock (at Yale) for their encouragement, helpful criticisms and suggestions in response to this work. The Wellcome Institute Library, London and the Cambridge University Library have assisted me in various ways during my research for this book. They have also kindly granted permissions for the use of illustrations. The British Library and the Bibliothèque Nationale have also been invaluable. I am indebted to Christ's College, Cambridge, the Maison des Sciences de l'Homme, Paris and the British Academy for their support in the development of this project. I am also grateful to the contributors at various meetings and conferences where I have presented some of the research now published here, especially to the participants at the seminar on War, Reconciliation and the State, held at King's College, Cambridge between 1989 and 1992.

A NOTE ON THE TEXT

Where the footnotes to quotations rendered in English refer to a foreign language source, the translation is mine. Italics are in the passage I quote unless otherwise stated. For the sake of clarity in cases where I cite untranslated work, I have generally provided the equivalent English title in the main text itself, whilst indicating the original in the accompanying reference and the bibliography. Where I quote consecutively from the same text, the second and all subsequent references are given simply as page numbers in brackets. Entries within the bibliography are divided into sections for pamphlets, articles and books. To distinguish pamphlets from books, the abbreviation [P] is used within footnotes. An essay derived from some of this research has already appeared as 'Pourquoi la guerre?' in the *Revue internationale d'histoire de la psychanalyse*, 5 (May 1992).

'...in diesen Tagen des ins ungeheuere gesteigerten Sterbens'

'...in these days of monstrously accelerated dying'
(Letter of 4 October 1914, in Rilke,
War Time Letters, p. 18); *Briefe*, p. 17

1 Introduction

Imagine two interlocking projects in the history of European culture and political thought: the one tracing the quest for a pure science of war; the other charting the dissolution of the belief that war can be reduced to laws or predictable patterns. Nowhere do two such trajectories of dream and nightmare intersect more strikingly than in the correspondence which took place between Einstein and Freud in relation to the question 'Why War?', precisely one hundred years after the appearance of the most famous of all rationalist war theories, Clausewitz's *On War* (1832). Einstein was invited by the League of Nations to write a public letter to the correspondent of his choice. He turned to the pioneer of psychoanalysis with the question of war. Nazism itself was not explicitly addressed by Einstein, but it evidently motivated and gave urgency to his inquiry. He requested new thinking and asked whether it was still possible to hope for an end to war once and for all. Freud responded with some weariness, unconvinced that such a dialogue would get very far or that he had anything new to say. Their exchange did indeed take place in the twilight of European aspirations for 'permanent peace'.

The dialogue provides a continuing point of reference and is the subject of my concluding discussion. It is placed, however, in a much wider context of thought on war and peace. This work draws on a range of primary sources on war, including popular fiction, philosophy, politics, historical writing, psychoanalysis, psychiatry and other human sciences. It explores why so many major nineteenth-century writers have justified war and even considered it rational and indispensable to social survival. Conversely, it traces the history of the nexus between war, madness and industry towards its apotheosis in the First World War crisis of shell-shock. The parallels and cross-over points between images of willed machines and automaton-like human beings preoccupy so many modern fictions and discussions. The 'doomsday machine', set in motion by the mad human agent but subsequently quite beyond recall, has been memorably portrayed in films like *Dr Strangelove*. Yet the language of nuclear fear echoes even as it transforms earlier responses to technical 'advance': from the terror of machine guns, tanks and bombers to the more abstracted horror of 'the war machine' itself.

In his essay of 1915, *Thoughts for the Times on War and Death*, Freud had noted the irony that scientific advances produced ever more destructive capability. War today is 'more bloody and more destructive than any war of other days, because of the enormously increased perfection of weapons of attack and defence'. Military conduct itself, however, remains 'at least as cruel, as implacable as any that preceded it'.[1] Technology changes; civilisation progresses, it seems; but primitive human aggression, the desire to inflict pitiless violence upon an enemy, apparently endures obstinately intact. The First World War confirms amongst other things for Freud the inexhaustible rage of the unconscious.

This study explores some of the ways change and continuity, technology and destructive power, rationality and derangement have been conceived in war writing between the age of Clausewitz and Freud. In many respects my choice of writers looks eclectic, to say the least. Clausewitz, De Quincey, Proudhon, Engels, Ruskin, Valéry and Freud, to name some of the major figures discussed, do not form, I readily confess, a self-evidently coherent group. Moreover to compound the problem these illustrious writers are linked to or juxtaposed with a larger cast of intellectual contributors and a plethora of popular stories of invasion, propaganda pamphlets and anthropological speculations about national identity. Yet there are also many surprising affiliations and echoes of model and metaphor across these disparate texts, written for such widely differing audiences, in separate circumstances, times and places. At the risk perhaps of underplaying the distinctiveness of national traditions of thought, indeed often precisely with the aim of displacing the question of 'national character', I seek to bring together this apparently eclectic collection of writings from the 1830s to the 1930s. The purpose is not to equate but to interrelate them.

Almost without exception, for instance, the writers I discuss are men, and often men for whom war evidently raises troubling questions of sexuality and gender, even though, at the same time, war is frequently said to resolve them. Sexualised imagery and argument run like a conspicuous thread through the pages that follow. The nation is itself often gendered as female; yet women apparently count little in the economy of the chivalrous battle. War is the fault of women. War is born from the womb of the state. War is the testing ground of virility but it disturbingly produces 'feminine' hysteria amongst the men. War is either very good or

1 *Thoughts for the Times on War and Death*, p. 278. For full details of references, see the bibliography.

very bad for racial reproduction. War is the mechanical human beast. War is the runaway train upon which interminable crimes of passion are committed. War erupts from a secret French steel tube thrusting out beneath the sea and henceforth England is no longer virginal.

'Scarcely a human being in the course of history has fallen to a woman's rifle,' declared Virginia Woolf in *Three Guineas* (1938), another of those remarkable 'letters on war' from the 1930s. It seems that the rationales for military conflict, as for private property and male exclusivity, have to be ceaselessly and wearyingly undone. According to Woolf, they *repeat* themselves compulsively, enduring long beyond any specific rational demolition of their arguments:

> It seems as if there were no progress in the human race, but only repetition. We can almost hear them if we listen singing the same old song, 'Here we go round the mulberry tree, the mulberry tree, the mulberry tree' and if we add, 'of property, of property, of property,' we shall fill in the rhyme without doing violence to the facts.[2]

I make no apology for quoting extensively from the primary literature I cite. Various theoretical discussions of the purposes of war and the function of enemies are set alongside chapters on the fear of Prussianism, the language of First World War propaganda and the invasion story between 1870 and 1914. It is as a history and a set of case studies of the imputed deep functions of war that I intend this essay. There are numerous precedents for the psycho-analytic recognition that something deeply symptomatic of the

---

2 Woolf, *Three Guineas*, p. 76. It is possible to read Woolf, at least at certain points of the essay, as endorsing a stark psychological and biological, as well as historical dichotomy between men and women, in their respective relations to war and Fascism. On the other hand, any adequate analysis of *Three Guineas* would need to stress, alongside the bi-polar gendered terms of guilt and innocence, implication and immunity the text offers, its powerful insistence that the current system of social and economic relations, above all for Woolf the system of education, upholds *for both sexes* given roles and mystifying identifications. It should be added here furthermore that any simple notion of psychological, textual or historical dichotomy, in which women could be placed straightforwardly on the political 'other side', anti-war, anti-Fascism, beyond the pull of the repressive and bellicose ideologies Woolf describes, has recently been called into question, in very different ways, from within feminist criticism and historical investigation. See Rose, *The Haunting of Sylvia Plath* and Koonz, *Mothers in the Fatherland*. How such work might then be used to reread, and rewrite, a study like Theweleit's *Male Fantasies* remains to be seen.

nature of war itself occurs in that 'pre-war phase' where the world
is ideologically divided up into two hostile and utterly incompatible
camps; whether they are called two states or just two apparently
opposite and irreconcilable positions, even perhaps the binary differentiation of people under militarist or pacifist headings, such
Manichaean views may themselves be part of the order of violence
to which they purport to be merely a response.

In recent years a variety of interesting new critical and historical
work on war has appeared which has left its mark on the present
study. Indeed even in the case of Paul Fussell's *The Great War and
Modern Memory* with which, as I explain later, I have more trouble
and with which in the end I feel it necessary to take issue, I want
to acknowledge the compelling and deeply evocative nature of
the undertaking. From explorations of war and gender in Klaus
Theweleit's disturbing work, *Male Fantasies* or in the collection
Behind the Lines, edited by Margaret Randolph Higonnet *et al.*, to
innovative new studies of First World War cultural history and
first-hand war experience, notably by Leed, Eksteins and Hynes, a
wealth of neglected primary material and valuable argument has
emerged.[3] Important surveys of nationalism and patriotism have
also been produced recently which raise challenging theoretical
questions as well as providing more localised case studies of the
terms of collective identification and the role of war. Again, I will
refer to some of these investigations explicitly later, but my book is
also in a broader sense a response to and a dialogue with such
literature.[4]

The present investigation, however, drawing both on my own
historical research and on the wider secondary literature, attempts
to provide new questions, connections and parameters for the consideration of war. It is neither a straightforward history of a particular war nor a theoretical statement about violent conflict in
general, but an attempt to draw out some of the key terms of the
discussion in the modern Western tradition in order to open up a
debate about both the language of war in the previous century and
the terms of our current thought. Clearly the war pronouncements

---

3   Leed, *No Man's Land*; Eksteins, *Rites of Spring*; Hynes, *A War Imagined*. For
    my appreciation of but reservations about Theweleit's *Male Fantasies*, see *The
    International Review of Psychoanalysis*, 15 (1988), 531–4.
4   Interesting and influential work in the last ten years includes Anderson,
    Imagined Communities, Hobsbawm, *Nations and Nationalism since 1780*,
    Nora (ed.), *Les Lieux de mémoire*, Kristeva, *Étrangers à nous-mêmes*, Samuel
    (ed.), *Patriotism*. I have tried to respond to the last of these in an essay review,
    'Patriotism', *Australian Journal of History and Politics*, 36 (1990), 413–19.

of all the major figures I discuss warrant further investigation. Each of these accounts could reasonably be the subject of specialist studies in its own right. My defence is not so much that adequate full-length investigations of individual texts and writers already exist (which is by no means always the case), but that my principle concern lies in the interrelations and reverberations between these writings; the aim is indeed to convey the reader into a kind of echo chamber of historical thought on war.

Again, it could be argued that the specific political, diplomatic and imperial histories of the nations in which these texts were produced demands far more differentiation than I provide here. My intention is not to force comparisons or equations between commentaries from different countries and times, nor to claim that all these ideas have had the same impact in each European political or military system, or in the relationship between the imperial powers and their colonies, but to enable a recognition of the remarkable cross-references of modern war thought, as it were a dictionary of overlapping received ideas across frontiers, whose connections go far beyond the question of explicit citation or direct authorial 'influence'. This approach risks perhaps a certain sense of 'overkill' (to draw upon a pertinent metaphor); it aims to scotch the reassuring idea that the endorsement of 'war for its own sake' (which became something of a militaristic corollary to the doctrines of late nineteenth-century aestheticism) was simply the view of a few mavericks and cranks, by providing an abundance of primary examples in support. Yet what follows is not intended as a tour through some interminable gallery of identical and ghastly historical portraits. If it suggests the enduring and overarching system of ideas and contradictions which we still so often reproduce, it also shows the pertinent, continuing and restless *dialogue* in the past about the difficulty of peace, the risk of modern 'progress' and the enduring psychological or political value of bloody military strife.

Clausewitz insisted that war theory must continually be re-conceived in the light of changing historical circumstances. Each age, he declared, must represent war to itself anew lest its thought petrify uselessly in the past and its armies come to grief in the present. The shock of Napoleon proved the inadequacy of any rigid adherence to the military 'old regime'. But in fact even for Clausewitz there was no simple escape from earlier ideas in his own reconceptualisations.

During the nineteenth century there were dramatic material transformations in the capacity for inter-state or civil war violence. There were new conditions of militarisation, but also, as I show, a widely shared vision in the later nineteenth century of the state itself, Prussia in particular, as a terrifyingly harsh war machine. As one of Clausewitz's commentators, Colonel Maude, was to put it in 1908, the warlike mentality in Prussia has been prepared 'by the training of a whole century' and now has 'become instinctive'.[5] What the image of Germany in general and Prussia in particular so often provided in both Britain and France was simultaneously an absolute marker of national differences *and* a contaminating force; *both* an opposite *and* a metaphor of dangerous infiltration. Further-more if Prussia was apparently a mindless war machine it was also an insidiously controlled state which understood perfectly the nature and the weakness of its enemy. In the decades before 1914, the proposed Channel Tunnel was to evoke in Britain powerful terrors – and extraordinary fictions – of continental invasion. Defence arguments were inextricably bound up with the wider social and political anxieties of the age. Transformations in the machinery of destruction were themselves inevitably caught up in and shaped by the language of war which, paradoxically, so often sought merely to provide an objective reflection of the changing material world.

In part my intention is to stress certain crucial shifts in cultural understanding, particularly in relation to the trauma of the Franco-Prussian War (1870–1). At the same time I seek to question some of the current conceptions of war and modernity; to suggest something of the complexity of the relations of past and present, permanence and novelty; to examine the notion of 1914 as divide between a prior age of innocence and twentieth-century bitter, ironic experience.

Between the time of Clausewitz and Freud, there have been many enduring dilemmas about how to think of war. Not least about the very conceptual consistency of the term. Is there a 'why' of war, which can be asked and answered in universal terms? Or does the intuition of modernity – whether glimpsed in Napoleon or the Battle of the Somme – mark an unbridgeable chasm with all previous 'wars' and all earlier reflections upon them?

The current study explores some of the crucial functions and crises attributed to war in and beyond the last century. It should be added that if it is evidently a *selection* of material from the last century, it still more obviously makes no attempt to deal with pre-nineteenth-century theorisations and debates, for instance all those 'Machiavellian moments' in the understanding of war described by

5   See Maude's introduction to Clausewitz's *On War*, p. 86. The page reference is to the 1986 Penguin edn.

Pocock.[6] Whilst my choice of material includes some of the more obvious and well-known commentators on war (Clausewitz, Engels, Treitschke and so forth), it also unearths, for instance, the little-known war essays of Proudhon, De Quincey and Ruskin. The content of this book has stemmed then from a mixture of factors; from strategy and serendipity. It moves between key figures in the existing 'canon' of war writing and esoteric, little-read titles; it is determined by the dictates of my overriding arguments and questions (which in turn, of course, have been reshaped with continuing work), as well as by the fortuitous leads resulting from the pursuit of obscure Victorian footnotes. I concentrate by and large upon two issues which, I contend, are negotiated and debated repeatedly in the writers discussed – war's supposedly anchoring role in the constitution of collective identity and in the advance of 'civilisations' on the one side; war's intrinsic capacity to undermine, disturb and exceed predictions about its nature on the other. I argue that these conceptions have indeed been deeply important aspects of the historical representation of war and that they warrant consideration now. I show how the figure of war is torn between such different discursive possibilities: it is seen to provide coherence, boundaries, meaning but also to erode the identity of the structures and forces that inaugurate it. War has been conceived as cohesive, purposive, stabilising, but elsewhere as the consequence of a mad and uncontrollable machine or machine age; as protean and slippery, amenable to many definitions over time, fragmented by its countless national and local forms.

For Clausewitz, war is always to be understood as subordinate to political will. That is an iron law. But it also slips out of control, threatening to become jubilantly and anarchically autonomous. It is willed, but all too prone to succumb to chance and accident. 'Friction' is itself a law. The very etymology of the word 'war', one might add, takes us back to the old high German *werra*, meaning confusion, discord, strife. The practice of war, Clausewitz contends, can be shown to undermine the consistency of thought and theory upon war. So often in the past, he shows, the conduct of new campaigns has revealed the redundancy of the models and theories contemporaneous with them. Indeed some would contend that such a discrepancy is axiomatic, above all today. Moreover a final nuclear war is arguably unthinkable and untheorisable in the unique sense that, whilst it can be anticipated according to the terms of our current language and memory, its actual occur-

6 See Pocock, *The Machiavellian Moment*, esp. pp. 176, 200–1, 402.

rence would overrun and obliterate the very possibility of its own retrospective representation. This book concerns itself not with the chronicle of individual battles, nor with the psychological impact of war on soldiers,[7] but rather with war as an idea, an abstraction, a supposed structural necessity; and also as an impossible subject, the subversive force in the account which seeks to master it.

Whilst Clausewitz is an obvious exception, much of the literature introduced here is written from the sidelines of international conflicts. The object is not to demonstrate that such writings invariably had an immediate practical efficacy, nor that they found their way straightforwardly into military handbooks. Some of the more baroque fictional and philosophical visions detailed in the pages that follow were not translated directly into the army manual; not part of the terms of reference of such well-thumbed works as *The Active Service Pocket Book*, which went through five editions in the Edwardian period.[8] But nor of course should we see military writing and teaching as separable from the wider social ethos and cultural debate by which it is affected, and which it also influences.

Military writing around the turn of the century displayed on occasion both the extraordinary anachronisms and the prescient anticipations which characteristically divided so much other pre-1914 literature. To read, for instance, the scribblings of the Commanding Officer 6th (Poona) Division, Major-General E.A.H. Alderson's *Lessons from 100 Notes Made in Peace and War* (1908) is to find the familiarly ludicrous Edwardian representation; war as hunting and cricket; a matter of common sense and individual initiative; the scene of smart uniforms and manly vigilance. Alderson offers Napoleonic maxims, literary quotations, Kiplingesque homilies to 'the glorious game of war'.[9]

On the other hand if we go to Lieutenant-Colonel Eugène Hennebert's *Military Art and Science* (1884) we find an evocation of industrialised conflict, an awareness of factory production for

7 On the impact of war upon its participants and the ritual ordering of the chaotic experiences of combat see Leed, *No Man's Land*. Leed deftly explores some of the metaphors of war, machinery, identity and historical division (with which in general terms I will also be concerned) as they figure in soldiers' own accounts of the meaning of the First World War.

8 See Lieutenant Stewart, *The Active Service Pocket Book* (1906); 5th edn 1912. The book ranges across the composition of brigades and divisions, 'humane' laws and 'civilized' customs in modern war, methods of defence, retreat and attack, terrains, first aid, sanitation, advice for military cyclists and on horse management, camp cooking, field engineering, the make-up and operation of the machine gun.

9 Alderson, *Lessons from 100 Notes Made in Peace and War*, p. 36.

1 'Le Grand Marteau-
Pilon du Creusot' (from
Eugène Hennebert, *L'Art
militaire et la science*, Paris,
1884).

2 'Effets de l'explosion
d'une torpille sèche ou
fougasse ordinaire', Passe
de Schipka, Balkans, 21
August 1877 (from Henne-
bert, *ibid.*).

war which dwarfed the individual human being (see figures 1 and 2). Here again we cannot divorce the military writer from wider investigations on war in literature, anthropology and philosophy. Hennebert's warning, *The Imminent War* (1890), draws specifically on the anthropological speculations about 'Prussianism' which I explore below.[10]

I focus on perceptions of the meaning of war, overviews of the function of conflict written by novelists, anthropologists, psychiatrists, poets, natural scientists, journalists and ruminative military officers which shared so many features in and beyond the nineteenth century even though professional soldiers have often derided the ignorance and naivety of their lay critics and supporters.

'The result, as you see, is not very fruitful when an unworldly theoretician is called in to advise on an urgent practical problem,' Freud wrote ruefully to Einstein about their discussion of war.[11] But then again it is not clear that any overall account of war *is* exactly 'fruitful', from whatever supposedly 'worldly' or 'unworldly' position it is written, nor that combatants are by definition closer to some universal truth of war.[12] In any event, the most productive accounts of war, it seems to me, are those which recognise, precisely, the unavoidable roughness of the outcome, the lacunae, the inconsistencies of the execution, the questionable nature of the very enterprise to encompass war in writing. The present work has no catch-all thesis to offer on war; rather it lays out a particular constellation of representations, thereby at least to call into ques-

---

10 Thus Hennebert quotes the incontrovertible 'scientific' evidence of the anthropologist Quatrefages (see Chapter 6) to the effect that the Prussians are born predators, quite separate from the Germans, owing their peculiar racial character to their 'Finnish' descent. Hennebert warns against the imprudent mood in France which led to the *Exposition Universelle* of 1889, showing off French industrial greatness, arousing once again the 'envy' of 'scavenging' nations. His exploration of the elaborate network of spies and traitors passing information from France to Germany is one instance of a wider climate of racial dread and excoriation, intense fears about miscegenation and the seepage of military information, in the decade of Dreyfus. See Hennebert, *La Guerre imminente*, pp. 11–12, 21 and *passim*.

11 See Freud, *Why War?*, p. 213.

12 In a recent essay in *Critical Inquiry*, Susan Schweik has explored and questioned the terms in which authenticity and authority are ascribed to the 'soldier poets' of the First World War. Those terms are in turn used to account for the effective marginalisation in many traditional historical and literary critical discussions of women's writing and experience of war; see Schweik, 'Writing war poetry like a woman'. Categories like 'combat', 'frontline' and 'home front' cannot be presupposed in the discussion, she argues, but themselves demand historical investigation and critical analysis.

tion some of the clichés of our current discourse, the platitudes which still have their purchase today, their numbing and exonerating effect. It aims to provoke a certain hesitation, a recognition of the sheer difficulty – something of the unease and the embarrassment felt by Freud when Einstein asked him for an answer. Thus we should at least retain some scepticism and draw upon a longer history of metaphors and myths about war and emotion, war and momentum when considering the recent pronouncement of Colonel Michael Dewar, Deputy Director of the International Institute for Strategic Studies:

> it is naive to criticise either Schwarzkopf or Bush for the slaughter, in particular that which occurred at Mutla Ridge, before the ceasefire was called. First warfare – particularly blitzkrieg – acquires a momentum of its own which is difficult, even impossible, to stop.
>
> Second, and most important (as unpalatable as it may be to armchair commentators), emotions are heightened in battle.[13]

Or to take another formulation of this point from an entirely different source:

> Battle is now nothing more than the autonomy, or automation, of the war machine, with its virtually undetectable 'smart' weapons such as the Exocet missile, the Beluga bomb, the Tigerfish torpedo, the 'Raygun Project' of lightning nuclear attack being studied by the Pentagon, the Doomsday machine . . .[14]

The 'unstoppable engine of war' has become something of a modern truism. As though in answer to the question 'Why War?', the answer finally might turn on the insatiable and irresistible drive of the 'military-industrial complex' – that ambiguous phrase crucially begging the question of human agency or responsibility. Or as though war today might not only involve the deployment of new technologies, but be *essentially* redefined by them: satellites, television, computers and video games after all have occupied centre stage in the representation and military news management of the Gulf War. Are we not so often caught in the fascination and massive distortion of this 'high-tech' image of conflict today – as though war is both decreed and exclusively played out by high-precision automata? But how new are such perceptions of war's own technological triumph and irresistible momentum?

13 Dewar, 'A defence of Mutla Ridge', p. 19.
14 Virilio, *War and Cinema*, p. 7.

If we turn to the pacifist C.E.M. Joad's last-ditch plea in 1939, also entitled *Why War?*, we find again this concern with the free-wheeling movement of conflicts. Whether the force of this momentum must be understood in relation to the war drive of the human psyche or the impetus of the war machinery itself is not immediately clear, but the coming fight is seen as a potentially uncontrollable spiral of events. Joad challenges the idea that the future war against Nazism is yet irresistible; but he recognises that once started it will blaze away ferociously; it moves beyond the powers of the human agency which set it alight: 'War, and in particular modern war, is like a forest fire; once it is started, none can set bound to the resultant conflagration.'[15]

War tends to unleash further rounds of conflict, he explains; thus the First World War and the Versailles Settlement contained the seeds of Nazism and further war. The First World War produced the Arab–Jewish struggle in Palestine as a result of the incompatible promises made by the British to the Jews and Arabs between 1914 and 1918.

Joad sees war as irrational; but what partially differentiates him from Victorian anti-war liberals like Cobden (whom I discuss in the following chapter), and what marks one of the historical and thematic parameters of my argument, is the recognition of a certain necessity operating in war; or at least the relegation of the notion that war is simply anachronistic (by virtue of its uneconomic wastefulness), a tawdry conspiracy of short-term vested interests, the corrupt and pathetic last gasp of an aristocratic order. War is cast as a quasi-autonomous energy system, by definition anarchic and virtually unstoppable once in motion.

War is shown to engulf reason; to overrun the warped 'logic' which began it. Above all, modern war is perceived to have exceeded Clausewitzean political rationales. Through this crisis of political incommensurability, Joad forlornly hopes, the catastrophe may yet double back, and become, phoenix-like, the bearer of new life and hope. Thus the bombing aeroplane, whilst representing the greatest single disaster in human history, 'may prove to be a blessing' (p. 50). Joad clutches at the straws of pre-1914 pacifist aspirations in suggesting that the reality of uncontrollable war has now become so acute that it would be unthinkable to declare one – it would be insane, or, to borrow from a later scenario, it would be MAD – Mutual Assured Destruction.

15 Joad, *Why War?*, p. 89; *cf.* Swanwick, *Frankenstein and his Monster.* *Aviation for World Service* [P.]: 'mankind is Frankenstein; science, especially the science of aviation is his monster' (p. 5).

Joad's *Why War?* rejects the argument that war is natural. Indeed his book notes how war has been spuriously justified on this basis across the ages. From Heraclitus to Nietzsche, from Sir Arthur Keith (coiner of the expression that war is 'nature's pruning hook') to the Prussian military writer Bernhardi, or Mussolini, Joad finds dubious cosmic and socio-biological defences of military conflict. In fact, however, he rejects the language of aggressive social evolution (which will be explored in Chapter 8) only to comfort himself with the equally questionable assumptions of a pacifist theory of eugenic development: 'Wars tend to destroy the best and most efficient members of a nation, and leave the feeble and inefficient to live. Thus, as a result of war, a nation becomes not robust but decadent' (p. 50). Christianity and pacifism, he adds, are 'not only sound morals but sound biology' (p. 79). Freud also had something to say on this subject, as we will see later.

In his essay, *The Moral Equivalent of War*, issued as a pamphlet by the American Association for International Conciliation in 1910, the philosopher William James sought to ground his own hopes of peace in a serious analysis of the impulse to go to war. It is a particularly fascinating discussion, which I want to set out here as a coda to this introduction; it provides an agenda and a condensed summary of many of the questions, models and assumptions that will be at issue in my investigation. The essay might serve as a useful preface to a reading of Freud on war. At the same time, as it is precisely the objective of this book to show, both the war writing of a James or a Freud can be situated in a much wider history of war thought. Like so many of the other figures I discuss, James insists that nineteenth-century utilitarian beliefs (where war's value would be measured in terms of some 'felicific calculus' of overall human pain and pleasure) are quite inadequate. War cannot be dropped or picked up at will, as though it is merely some external tool or poison. The internal and the external aspect of war are difficult to separate. Moreover *The Moral Equivalent of War* both alludes to and manifests a problem about the relationship between language and violence. Conflict is not simply an external object of the discussion. One is almost invited to be 'in conflict' in the process of reading James, caught between the startlingly prescient insights and the entrenched taken-for-granted positions we are offered. The inevitability of a political and ethical dilemma about the in-dispensability of war, or at least the integrative function of enemies in holding the subject together, is itself the ambiguous 'message' to emerge. There is a bewildering friction between the flexible and the intransigent elements of this discussion. James will eventually express his scepticism about war rationales, and yet his own argument is founded on various absolutely basic assumptions, for instance about the biological foundations of the warlike instinct. He problematically equates 'violence' (in terms of individual actions in human society and in the animal world) and 'war'. But it is not self-evident that we should run these two terms together. The state cannot simply be extrapolated from the 'sum' of either individual desires or instinctual actions. War is primarily, as *The Oxford English Dictionary* puts it, 'hostile contention by means of armed forces, carried on between nations, states, or rulers, or between parties in the same nation or state'. For James the violent individual impulse underlies the decision of states to go to war. It has been bred into the modern human being, in Lamarckian fashion, through the inheritance of the acquired characteristic of cruelty. We are in that sense apparently born as warriors because of the militarism of the past: 'We inherit the warlike type; and for most of the capacities of heroism that the human race is full of we have to thank this cruel history.'[16]

James insists that the drive to commit violence and aggression lies deep in the body, the mind and the history of the species: 'Our ancestors have bred pugnacity into our bone and marrow, and thousands of years of peace won't breed it out of us. The popular imagination fairly fattens on the thought of wars' (p. 6). In contemporary society, however, it has become far more difficult to admit the love of war outright. The pleasure of attack is rationalised as defence, the quest for 'pure loot and mastery' is hidden beneath 'pretexts' which erroneously attribute such motives and forces exclusively to the enemy. In this climate, the advocacy of 'peace' is highly suspect, indeed often it is simply the pious rhetorical veil beneath which war calculations continue to be made.

James challenges the distinction between the two terms, 'war and peace'; rather than being opposites, they describe in modern life simply the gap between the implicit and the explicit mode of violent conflict. The temporality of wars thus requires reinterpretation. Battles do not necessarily begin when the first shot is fired. The relationship between language and military deeds must be reconceived. Words, ideas, images constitute the discursive support for military conflict; they should be understood not as though they are mere froth without consequences, but as crucial aspects of the destructive reality of violent conflict itself. The preparations for military campaigns, whether at the level of propaganda or of

16   James, *The Moral Equivalent of War* [P.], p. 6.

armaments factories, must be understood as integrally caught up in the conflict, not merely its precursor, let alone its antidote:

> Peace in military mouths to-day is a synonym for 'war expected.' The word has become a pure provocative, and no government wishing peace sincerely should allow it ever to be printed in a newspaper. Every up-to-date Dictionary should say that 'peace' and 'war' mean the same thing, now *in posse*, now *in actu*. It may even reasonably be said that the intensely sharp competitive *preparation* for war by the nations is *the real war*, permanent, unceasing; and that the battles are only a sort of public verification of the mastery gained during the 'peace'-interval. (pp. 6–7)

If the preparation for war *is* the real war, it follows that the relationship between war and peace, and between language and violence, is vexed indeed. But it would be far too hopeful to imagine that one can succeed in averting war by simply calling for an end to its preparation. War, James insists, is fundamental; it is 'the romance of history' (p. 8). Pacifism too often pitches its message at the level of a reasoning voice which is unable to provide the romantic non-utilitarian significance for which the human being strives.

One of the enduring themes across the nineteenth-century war literature I cite is that war constitutes in its essence a transcendence of all petty calculations and self-serving motives. Like art as understood in so much nineteenth-century theory, war is not to be viewed as a means to an end, but as an end in itself. War, it is suggested, is capable of defining precisely what it is to be human, because it involves giving up the supreme 'self-interest', life itself. It is in that sense the prerogative of risking death which defines warring man as more than an animal. In this view, war is a necessity not so much because the biological realm of 'nature' itself is red in tooth and claw, but because it captures the irreducible particularity of the human spirit. Set against such a philosophy which recognises the deep-defining function of war – its aesthetic, ethical and psychological purposes, its sheer human meaningfulness – James suggests that the conventional intellectual cupboard of the pacifist is bare. It cannot compete with the inspiring 'mystical' impulse manifest in militarist writing:

> [War's] 'horrors' are a cheap price to pay for rescue from the only alternative supposed, of a world of clerks and teachers, of co-education and zoophily, of 'consumers' leagues' and 'associated charities,' of industrialism unlimited, and feminism unabashed.

No scorn, no hardness, no valour any more! Fie upon such a cattleyard of a planet! (p. 8)

Ambiguously poised between ironic paraphrase of the discourse of the militarist and philosophical endorsement of the actual primacy of war (which entails too a denigration of 'feminism' as part of the dreary and impoverished world of 'soft' peacefulness), the passage reproduces the ambivalence which James insists that we must all share. The militarist is clear that war is indispensable; if it ever stopped, 'we should have to reinvent it, on this view, to redeem life from flat degeneration' (p. 8). The opposition between militarist and pacifist is itself in question since, 'So far as the central essence of this feeling goes, no healthy minded person, it seems to me, can help to some degree partaking of it. Militarism is the great preserver of our ideals of hardihood, and human life with no use for hardihood would be contemptible' (p. 8). Accepting that argument, it follows that the peace-loving James must find a 'moral equivalent' of war, if a hopeless and pathetic social lethargy is to be avoided.

But there is another sense in which war and the words of war *are* keepers of the peace, since they help free the human subject from the awareness of internal conflict. A world of peaceful pleasure with no external conflict would be terrifying indeed. It can be summed up, James suggests, by acknowledging:

> that mankind was nursed in pain and fear, and that the transition to a 'pleasure-economy' may be fatal to a being wielding no powers of defense against its disintegrative influences. If we speak of the *fear of emancipation* from the *fear-regime*, we put the whole situation into a single phrase; fear regarding ourselves now taking the place of the ancient fear of the enemy. (pp. 11–12)

War has so much to offer. How can we give up its charm, its speed, its thrills, its sense of tragedy in the name of a gradual and insipid peaceful evolution? There will always be a deep unwillingness to see 'the supreme theatre of human strenuousness closed, and the splendid military aptitudes of men doomed to keep always in a state of latency and never show themselves in action' (p. 12). Pacifist talk of war's horror and expense misses the point that both those factors are precisely what makes it worthwhile. In short: 'Pacifists ought to enter more deeply into the esthetical and ethical point of view of their opponents' (pp. 12–13).

James seeks to conciliate between the pacifist and the militarist by showing them that they are not simply on opposite sides, that indeed *all* 'healthy' people share a certain fascination with war, a certain inevitable complicity which is both psychological and biological. With its quasi-Nietzschean contempt for femininity, impotence, softness and peace, the 'socialist' and 'pacifist' James wants to transpose all that is good about war into the economy of a permanent peace. That is the only way to realise non-violence, he claims:

> A permanently successful peace-economy cannot be a simple pleasure-economy. In the more or less socialistic future towards which mankind seems drifting we must still subject ourselves collectively to those severities which answer to our real position upon this only partly hospitable globe. We must make new energies and hardihoods continue the manliness to which the military mind so faithfully clings. Martial virtues must be the cement; intrepidity, contempt of softness, surrender of private interest; obedience to command, must still remain the rock upon which states are built – unless, indeed, we wish for dangerous reactions against commonwealths fit only for contempt, and liable to invite attack whenever a centre of crystallization for military-minded enterprise gets formed anywhere in their neighbourhood. (p. 15)

So James rejects the notion of the inevitability of the 'war function' as fatalistic nonsense, but the 'competitive passion' which underpins it *is* our fate. Yet rather startlingly he proceeds to reassure us that an end to war is itself inevitable since technology has reached such a level of destructiveness that its use has now become 'absurd and impossible': 'And when whole nations are the armies, and the science of destruction vies in intellectual refinement with the sciences of production, I see that war becomes absurd and impossible from its own monstrosity' (pp. 14–15). James warns that the ensuing peace will be terribly effete unless the spirit of war is co-opted. Instead of actual war, human beings must do their mining, cleaning, building, and so forth with a new martial vigour, a new culture of 'toughness' which captures and displaces the glories of the warrior spirit.

With its strange ambiguities of argument, its doubling and tripling back upon itself, its admirable refusal of Manichaean distinctions, its suspicion of the naive basis of both conventional pacifism and militarism, but also its own descent into the very orthodoxies of race, gender and biology, in short into the propaganda it sets out to challenge, James's argument is absolutely exemplary of the discussion to follow. The dubious assumption of a dramatic racial gulf between 'yellow' and 'white' peoples coexists, for instance, with James's recognition that propaganda is needed to produce or at least bolster the sense of the 'yellow' Asian people as some kind of immutable enemy. War is rejected by James but in the name of a kind of phallic index of culture, an endorsement of a militarised peace which must remain our biological and cultural inheritance: 'there is no reason to think that women can no longer be the mothers of Napoleonic or Alexandrian characters' (p. 9). At the same time James interestingly questions the divide between 'peace' and 'war' in ways that might usefully lead us to question the very periodisation of the 1914–18 disaster.

My intention is to examine the perception and periodisation of war in and beyond the nineteenth century; to make links across disparate intellectual enterprises, fictions and sciences on war; to explore a world of representations which sets in play a 'common sense' and a debate about wars, states and states of minds which is still very much contemporary for us, a language and vision which we still share, or at least upon which we still draw. Many of the writings surveyed here have profound resonances – much to say to us about war and peace – but they also speak their history, their place in wider networks of argument and representation. They speak a language which, by definition, they neither own nor simply invent, a language which they take up, but which they also may disturb and shift. A further central purpose of this study will be to demonstrate the wide-ranging provenance of and varied nuances within the contention that war is functionally necessary. This idea was often given idiosyncratic national inflections and had specific ramifications in different contexts; but it had no single intellectual home, even though the Germanic or Prussian monopoly of war lust has often been a crucial part of the story Europe has told itself.

2 Cobden's Critique of War

Before turning to the war writers who form the central focus of this study, I sketch out, for contrast and comparison, Richard Cobden's arguments for peace. His work forms part of a much wider history of European rationales for non-violence in international relations which can be consulted elsewhere.[1]

In the course of this impassioned criticism, Cobden raises a number of issues about the constitutive role of fantasy in the construction of nationhood and in the formation of enemies, which are highly pertinent to my discussion. Fantastic scenarios of danger, the imaginary horrors of invasion and desecration, he warns, are difficult to extricate from genuine external menaces. There is an abiding problem, he suggests, in distinguishing the potential real violence of an object from the projections of the subject. Cobden felt that the political culture in which he lived was often haunted by phantoms rather than actual external danger, just as it apparently had been in the 1790s when Burke's 'frenzied imagination' did so much damage.[2] He weaves together his diagnoses of the psychopathology of invasion (Burke's monomania, his dagger, his dread of French infection), with his economic and religious objections to military answers. War, we are told, is usually wrong for both pragmatic and high moral reasons. He declares aggressive wars of conquest totally unacceptable and seeks to analyse British chicanery, meanness and treachery in Asia, the spurious rationales 'got up' by the imperial power to justify the vindictive and brutal exercise of its own might.[3]

Manufacturer, leader of the anti-corn law league, MP and *laissez-faire* liberal champion, Cobden spoke indefatigably on the subjects

1 See, for instance, Silberner, *The Problem of War in Nineteenth Century Economic Thought*; Brock, *Pacifism in Europe to 1914*; Nicholls, 'Richard Cobden and the International Peace Congress Movement'.
2 Cobden suggests that Burke's xenophobic stance in parliamentary debate in the 1790s stemmed from 'monomania': 'At length, on the discussion of the Alien Bill [on 28 December 1792], Burke's powers of reason and judgement seemed to be entirely overborne by a frenzied imagination. Drawing forth a dagger and brandishing it in the air, he cast it with great vehemence of action on the floor. "It is my object," said he, "to keep the French infection from this country; their principles from our minds, and their daggers from our hearts!"'
3 See, for instance, 'China War', in *Speeches*, vol. II.

of war and peace from the 1830s to the 1860s. His views were widely known, circulated via his speeches both in and outside the House of Commons, newspaper reports and pamphlets. His elaborately drawn parallel between the apparently groundless fears of the 1790s and the 1850s, '1793 and 1853', for instance, was written as three letters to an anonymous correspondent, 'the Reverend ———' but was in fact published in *The Times* and the *Manchester Examiner*. A pamphlet edition running into tens of thousands was also distributed.

Whatever the enduring refrain of older themes and arguments, nineteenth-century writers on war and peace often felt themselves to be – and in many respects were – responding to discontinuity. Cobden himself was to remark on both the novelty of the modern war, and on the new grounds for peace. Fighting is always a failure of reason, he insists, but in the modern age, war is nothing less than a catastrophic eclipse of sense, a bestial and mechanical descent into anarchy. As he warned in 1861, the year of the outbreak of the American Civil War, military conflict, once started, inevitably expends itself to the point of exhaustion:

> From the moment the first shot is fired, or the first blow is struck in a dispute, then farewell to all reason and argument; you might as well attempt to reason with mad dogs as with men when they have begun to spill each other's blood in mortal combat.... when a war is once commenced, it will only be by the exhaustion of one party that a termination will be arrived at.[4]

Cobden denounced scaremongering generals, insisting on the chimerical nature, for instance, of the Russian threat which was used to justify the Crimean War (1854–6). Consistent with his own earlier expressed views in the 1830s that alarmism about Russia was unjustified, Cobden continued to call for non-interference. He lamented the disastrous consequences of defence panic in wasting resources and in diverting youth (particularly of the middle classes) from civil pursuits to military exercises.[5] He was particularly scathing about the 'infantine alarms' of the aged Duke of Wellington. During the Great Exhibition of 1851 the octogenarian Duke was 'haunted by terrors' of a French invasion and refused the household regiments their usual summer retreat, surrounding the metropolis with troops. As Cobden complains, 'Everything in [France] is viewed by us through a distorted and prejudiced medium'.[6] The

4 Cobden, *Speeches*, vol. II, p. 314.
5 See Cobden, *Political Writings*, vol. II, p. 423.
6 *Ibid.*, vol. I, pp. 446, 452.

Duke had in effect become enslaved by inner compulsions, reduced to little more than a machine, or to put it another way, he was a patient who had lost his capacity of free thought:

> In his public capacity he never seemed to ask himself – what *ought* I to do? But what *must* I do? This principle of subordination, which is the very essence of military discipline, is at the same time the weak part and blot of the system. It deprives us of the man, and gives us instead a machine; and not a self-acting machine, but one requiring power of some description to move it. (p. 370)

We will repeatedly return to this envisaged relationship between war and machinery, war as 'machine mentality', in the present study. From the fiction of Samuel Butler and the social scientific investigations of Prussian national character after 1870, through the robot drama of Karel Čapek, *R.U.R.*, to Reich's diagnosis of the nature of Fascism in the 1930s, we will come back again and again to a certain crossroads of representation: the psyche potentially reduced to a machine-like state and the machine imbued with excessive destructive power, even a kind of sexual energy. In Cobden too, the question of ultimate control over or enslavement to war beliefs is answered in various ways. The notion of 'instinct', or as he puts it 'pugnacious propensity', is brought in, but its origins or current bearing remains ambiguous. Sometimes he detects a kind of thoughtless reflex action motivating the views of militarists; at other times he perceives a cynical and fully conscious plot.

Cobden is convinced that in the 1790s the pretence of national danger had masked the real objective: to destroy the system of government and ideas heralded by the Revolution.[7] It is not that he is an enthusiast for the French system, he reassures his audience; but hypocrisy must be rooted out of his government. Britain purports to behave in a pacific manner, whilst in fact:

> We have been the most combative and aggressive community that has existed since the days of the Roman dominion. Since the Revolution of 1688 we have expended more than fifteen hundred millions of money upon wars, not one of which has been upon our own shores, or in defence of our hearths and homes... this pugnacious propensity has been invariably recognized by those who have studied our national character.[8]

7 *Ibid.*, pp. 417–18.
8 *Ibid.*, p. 492. He cites the popularity of 'mad-cap' Richard, Henry of Agincourt and the belligerent Chatham, to say nothing of the erecting of monuments of

Just as trade should function 'freely', unencumbered by tariffs and any other unnecessary government obstacle, so diplomacy should be based on the principle of non-interference. Cobden insists that whatever may be the liberal's sympathy for oppressed nationalities, their claim can never rightly be supported by the external force of arms.[9] Where there is no such genuine national awakening at stake, the argument for military interference and aid is even more spurious, as Cobden was vociferously to protest with regard to Britain's support for what he saw as a ghastly non-Christian Turkey, during the Crimean War. He dissects at length the folly of the campaign, the drastic miscalculations of the strength of Russian forces, the poverty of information, the naive hope that Sebastopol would fall quickly, the general confusion and absurdity of the war aims.[10]

To back Turkey against Russia is in his view to support barbarism against Christianity (albeit of a distasteful kind).[11] Whilst Russia shows encouraging signs of both economic and moral progress, Turkey is the scene of horror, ignorance and loathsomeness.[12] It is a land desecrated by a fundamental and inalienable 'Ottoman violence', which is not only morally vile and diseased (whole provinces 'little better than nurseries of the plague!'), but economically ruinous.[13]

The best attitude to the disputes of other peoples is a principled neutrality, helping neither side rather than both. Even in the

warriors, the myriad memorials of battles in the names of bridges and streets as so many examples of this bellicose national character. The English should try to look less at the threatening posture of the French, more at their own; see pp. 453, 493. On the extent to which eighteenth-century Britain had indeed become 'a war machine', see Brewer, *The Sinews of Power*.

9   See Cobden, *Speeches*, vol. II, p. 6.

10   See Cobden, *Political Writings*, vol. II, pp. 110, 140.

11   See Cobden, *Speeches*, vol. II, p. 6. *Cf.* the appendix to 'Russia' [1836], 'Extracts from various writers illustrative of the condition of Turkey', which focused on various war atrocities, most horribly the supposed propensity of the Turks to take fragments of the human body as trophies. In this passage, describing streets piled high with chins, ears, noses and other features of the slain, the Cobdenite Orientalist tirade reaches its delirious climax: 'The features, growing soft by putridity, continually attached themselves to their feet, and frequently a man went off with a lip or a chin sticking to his slippers, which were fringed with human beard, as if they were lined with fur' (*Political Writings*, vol. I, p. 341). Cobden's views on war and peace differed in many respects from Thomas De Quincey's (discussed in Chapter 6), but shared something of the same fear of Eastern contamination. See John Barrell's compelling recent study, *The Infection of Thomas De Quincey*.

12   See Cobden, *Political Writings*, vol. II, p. 118.

13   *Ibid.*, vol. I, p. 22, vol. II, pp. 330–1.

American Civil War, Cobden still called for non-interference, although here he seemed to waver somewhat.[14] For some other peace writers the American Civil War, together with the Prussian wars of the 1860s, effectively shattered the assured Cobdenite link between industrialisation and peace, progress and the transcendence of militarism.

In Cobden's view war remained fundamentally atavistic and regressive; he exhorted his audience to renounce it. Peace is always preferable – although sometimes he adds the caveat where the 'honour' of the country can be preserved without war.[15] In place of military conflict, he proposes the establishment of a system of arbitration for international disputes. Again and again Cobden rests his argument on human nature, or more specifically on the deep-seated pacific mercantilism of the European people at large. Thus whatever Louis Napoleon's ambitions may be, he will be constrained by the good sense of the ordinary French individual and even the good will of the ordinary soldier. War is never desired by 'the people', but only by politicians and military men whose ambition and cupidity are fired by prospects of advancement or profit.[16]

Historically, Cobden suggests, the blame for bellicosity rests with the aristocracy. Their value system, their old regime of war romanticism, persists across the nineteenth century, despite the partial victory of the 'pacific' middle class. Native energy, daring and enterprise were 'perverted' in the past by an aristocracy desirous of 'constantly recurring wars'.[17] The crucial task of the moral politician is always to pull the people away from the cynical advocates of war, who have triumphed all too often since 1789: 'The people of this country must first be taught to separate themselves in feeling and sympathy from the authors of the late war, which was undertaken to put down principles of freedom.'[18] Indeed it is Cobden's premiss that 'no government has the right to plunge its people into hostilities, except in defence of their own national honour or interests'.[19] He exhorts his country to abandon its vain

14 See Cobden's speech in the House of Commons on the American War, 24 April 1863, *Speeches*, vol. II.
15 On Cobden's collaborator John Bright's view that Lincoln's administration was justified in fighting against the attempted secession by the South and more generally on the debate amongst Quakers about whether to make exceptions to the rule of war opposition, see Brock, *The Roots of War Resistance*, p. 56.
16 See Cobden, *Speeches*, vol. I, p. x.
17 Cobden, *Political Writings*, vol. I, p. 493.
18 *Ibid.*, p. 488.
19 Cobden, 'England, Ireland and America', *ibid.*, p. 9.

omnipotent pursuit of empire. As an island, Britain is deemed particularly safe, and thus particularly perverse and financially foolish in seeking to extend itself:

> with an invulnerable island for our territory, more secure against foreign molestation than is any part of North America, we magnanimously disdain to avail ourselves of the privileges which nature offers to us, but cross the ocean, in quest of quadripartite treaties or quintuple alliances, and probably, to leave our own good name in pledge for the debts of the poorer members of such confederacies. To the same spirit of overweening national importance, may in great part be traced the ruinous wars and yet more ruinous subsidies of our past history. Who does not now see that to have shut ourselves in our own ocean fastness, and to have guarded its shores and its commerce by our fleets, was the line of policy we ought never to have departed from – and who is there that is not now *feeling*, in the burthen of our taxation, the dismal error of our departure from this rule during the last war?[20]

In Cobden's view war is always deleterious, inimical to the interests of capitalism. As he put it in 1836:

> On behalf of the trading world, an indissoluble alliance is proclaimed with the cause of peace; and, if the unnatural union be again attempted, of that daughter of Peace, Commerce, whose path has ever been strewed with the choicest gifts of religion, civilization, and the arts, with the demon of carnage, War, loaded with the maledictions of widows and orphans, reeking with the blood of thousands of millions of victims, with feet fresh from the smoking ruins of cities, whose ears delight in the groans of the dying, and whose eyes love to gloat upon the dead: – if such an unholy union be hereafter proposed, as the humblest of the votaries of that commerce which is destined to regenerate and unite the whole world – we will forbid the bans.[21]

To fight a war in order to resist an embargo is absurd, he argues, since the pressure towards trade (which is to say innately self-interested commercial human nature) is virtually irresistible, regard-

---

20 *Ibid.*
21 'Russia', *ibid.*, pp. 323–4. Cobden goes on to insist that 'no class or calling, of whatever rank in society, has ever derived substantial or permanent advantage from war'. Admittedly the agriculturalist might gain some temporary advantage, but 'war is at best an intermittent fever; and the cure or death of the patient must at some time follow' (p. 324).

less of government intention. For this reason, Cobden proceeds, Napoleon's blockade, decreed in 1807, had a very small effect indeed in diminishing Britain's trade.[22] His account does not look in any detail at the ways in which war might perhaps stimulate industrial production in certain areas, focusing instead on the ways in which war fails to impede a pacific trade process occurring anyway.

These balances of trade, Cobden contends, must be taken into account when looking at our relationship with other countries, for instance Russia and Turkey. He stresses the large expansion in Anglo-Russian commerce and the small scale of business with Turkey. The possibility that Russia might threaten Britain's Indian possessions is rejected as illusory. The Tsar will be far too busy and too drained by dealings with Turkey to think of further expansion. In short, fighting Russia on the trade question is stupid.

But in any case not only limited wars in defence of far-flung colonies but the very existence of formal British colonialism is ruinous. Cobden refers to Adam Smith's doubts some sixty years earlier about the wisdom and profitability of such territorial power, and reiterates it. Think of Spain, which 'lies at this moment, a miserable spectacle of a nation whose own natural greatness has been immolated on the shrine of transatlantic ambition. May not some future historian possibly be found recording a similar epitaph on the tomb of Britain?' (vol. I, p. 25).

One of colonialism's central evils, he warns, is to distort the market, preventing the free circulation of goods. Where 'withering protection' (p. 28) prevents such natural, smooth economic respiration, degeneration inevitably follows. Free circulation is by definition natural and spiritually right, a universally valid principle which needs to be communicated far and wide. Free trade is nothing less than 'the international law of the Almighty'.[23] Capital itself is like a natural element which spreads itself equitably across the globe, except where it is artificially impeded: 'Capital, like water, tends continually to a level; and, if any great and unnatural inequality is found to exist in its distribution over the surface of a community, as is the case in this United Kingdom, the cause must, in all probability, be sought for in the errors or violence of a mistaken legislation' (p. 91). New journals of free trade should be established 'for the common object of enlightening the world' (p. 32). Indeed in this

---

22 In speaking of the folly of the Napoleonic Wars, Cobden drew reverentially on the name of Fox; see *Political Writings*, vol. II, p. 110.
23 *Ibid.*

rhapsody it is impossible to exaggerate the centrality of commerce to the future well-being of all:

> in the present day, commerce is the grand panacea, which, like a beneficent medical discovery, will serve to inoculate with the healthy and saving taste for civilization all the nations of the world. . . . Not a bale of merchandise leaves our shores, but it bears the seeds of intelligence and fruitful thought to the members of some less enlightened community; not a merchant visits our seats of manufacturing industry, but he returns to his own country the missionary of freedom, peace, and good government – whilst our steam boats, that now visit every port of Europe, [and] our miraculous railroads, that are the talk of all nations, are the advertisements and vouchers for the value of our enlightened institutions. (p. 45)

Colonialism is quite simply unnecessary: 'Provided our manufactures be cheaper than those of our rivals, we shall command the custom of these colonies by the same motives of self-interest' (p. 30). Cobden cites the West Indies and Canada as examples of places where this circulation is damagingly distorted; and the United States as an example of an ex-colonial state with which Britain now enjoys burgeoning trade.

The railway, the factory and the steamboat are all contributors to peaceful commerce and development, not conceived here primarily as destructive instruments of war. For Cobden, in the 1830s and beyond, war is quite simply historically retrograde. The Napoleonic wars led Europe backwards: 'all the nations of the continent fell back again into their previous state of political servitude, and from which they have, ever since the peace, been *qualifying* to rescue themselves, by the gradual process of intellectual advancement' (p. 45).

There is a certain paradox, however. One of Cobden's methods of reassuring his audience about the lack of real international danger, for instance from Russia, depends upon the existence of the advances in weaponry which he simultaneously denounces. Less 'civilised' peoples are now faced by insuperable military obstacles to whatever aggressive territorial inclinations they are imputed to possess. Military technology is thus for Cobden at once a scandal and a safeguard:

> the application of the power of chemistry to the purposes of war furnishes the best safeguard against the future triumph of savage hordes over civilized communities. Gunpowder has for ever set a barrier against the irruption of barbarians into

western Europe. War, without artillery and musketry, is no longer possible; and these cannot be procured by such people as form the great mass of inhabitants of Russia. (p. 339)

In a hypothetical contest of the old and the new, 50,000 modern-day Prussians could now defeat the vast and once terrifying hordes of yesteryear: 'Henceforth, therefore, war is not merely an affair of men, but of men, material, and money' (ibid.). Yet Cobden abhors the peacetime establishment of formidable stores of weaponry. He anxiously calculates the inexorable rise in the quantities of gunpowder, ball cartridges, cannon, cannonballs, shells and sandbags, and the rapid conversion of scientific discovery into new military technology. As he warns in 1849, peacetime becomes preparation for wartime:

> What are we to deduce from these facts? That instead of making the progress of civilization subservient to the welfare of mankind – instead of making the arts of civilization available for increasing the enjoyments of life – you are constantly bringing these improvements in science to bear upon the deadly contrivances of war, and thus are making the arts of peace and the discoveries of science contribute to the barbarism of the age.[24]

An unending spiral of accumulation has been set in motion, whose end is dark and unknowable: 'But I wish to know where this system is to end. . . . I see no limit to the increase of our armaments under the existing system. . . . I see no necessary or logical end to the increase of our establishments' (p. 169). The fatalistic acceptance of war – the taken-for-granted view of war one finds for instance in Clausewitz – was rejected. Palmerstonian adventurism was to be resisted at all costs. War should be criticised and strenuously opposed, he urged. Yet the whole of modern Europe, he lamented by the 1860s, 'has almost degenerated into a military barracks' and 'the greater the preparation the more imminent is the risk of collision' (pp. 457, 469). Cobden intimates and then recoils from the possibility that war might have become structural, built into the system, intrinsic to modernity. To view war as politics by other means, or worse still industry under a different name, was anathema.

24 'Foreign Policy', in Speeches, vol. II, p. 167.

3 Clausewitz and Friction

Clausewitz's *On War* (1832) sets in play a nineteenth-century dilemma about the nature of war in the post-Napoleonic world. In part this was a division about which Napoleon to represent – the Napoleon that shattered the old regime of war, or the Napoleon that came to grief in Russia in 1812. But it was also more abstractly and phantasmatically a dichotomy in the conception of war as such which recurs in different contexts. War as political instrument versus war as anarchy of the machine; each representation had multiple versions. This chapter explores the plurality of such resonances in Clausewitz's extraordinary treatise.

In an obvious sense, of course, Carl von Clausewitz (1780–1831) was not an 'armchair' commentator on war. A Prussian officer who served in various active military roles during the Napoleonic wars and who was captured during the Jena campaign, Clausewitz aimed to reflect his practical experience and observation, insisting upon the uselessness of mere abstraction. After his release, Clausewitz soldiered on against Napoleon, offering his services to the Tsar and playing a part in the Moscow campaigns of 1812 and 1813. Back in Prussian service, he spent the post-war years as director of the Military Academy in Berlin. His major writings on war were published posthumously by his wife.

But if Clausewitz's life was militarily active and his insistence was on practical experience, his work was just as evidently caught up in wider intellectual, philosophical and historical developments. Like Hegel, for instance, Clausewitz saw the advent of Napoleon, and in particular his victory over the German states, as a lesson of world historical significance.

In Hegel's *Lectures on the Philosophy of World History* (1822–30), the death of each nation is shown to contribute to the life of another greater one: 'It then serves as material for a higher principle.'[1] In the Napoleonic epoch, the fragmented German political system, a miscellany of kingdoms, principalities, duchies, bishoprics and free imperial cities, failed because, from the world historical point of view, it was already stagnant and dead, unable to gather itself into a vital unity.[2] Its time had passed. Clausewitz

---

1 *Philosophy of World History*, p. 60.
2 See Avineri, *Hegel's Theory of the Modern State*, p. 36.

makes a similar point in *On War* when he complains of how the Prussian Generals failed to adapt their theories to the new practices of Napoleon:

> When in the year 1806 the Prussian Generals, Prince Louis at Saalfeld, Tauentzien on the Dornberg near Jena, Grawert before and Rüchel behind Kappellendorf, all threw themselves into the open jaws of destruction in the oblique order of Frederick the Great and managed to ruin Hohenlohe's Army in a way that no Army was ever ruined, even on the field of battle, all this was done through a manner which had outlived its day, together with the most downright stupidity to which methodicism ever led.[3]

Napoleon broke the rules of an old 'civilised' warfare, revealing the bankruptcy of the codes that had hitherto governed military thought. He ushered in a new era of total war, and consequently a new age of theory. 'Theory is indebted to the last Wars,' declares Clausewitz, for revealing a new reality: 'the destructive force of the element set free' (p. 370). Modern theory, he argues, has explicitly to recognise its own historical implication – its methods and principles must necessarily be historicist. Since each period 'has had its own peculiar forms of War', so each period 'would, therefore, also keep its own theory of War' (p. 387). Again the affinity between Hegel and Clausewitz is striking.

History in the form of the French Revolution and the Napoleonic army had taught Prussia the hopelessness of its own adherence to eighteenth-century 'aristocratic' warfare in the face of its dramatic defeat in 1806.[4] The small professional army had to be displaced by a modern mass army. The military world had changed; the need for reform amongst the armies of Napoleon's opponents became overwhelming. Whilst there were setbacks to such reform, and indeed in Prussia the king regretted some of the reformist concessions he made, nevertheless the Defence Law of 3 September 1814 declared

---

3 *On War*, pp. 209–10. Citations, again, are from the widely available abridged Penguin edition. Reference should also be made, however, to the text as edited and translated by Howard and Paret.

4 On the Military Reorganisation Committee set up in Prussia in July 1807 under the leadership of Gerhard von Scharnhorst which reformed officer schools and threw them open to the middle classes as well as to the nobility, see Howard, *The Franco-Prussian War*, p. 11. The *Landwehr* was established in 1813 as a citizen's militia. Run and armed separately from the main forces, it fought alongside regular forces and contributed to the defeat of Napoleon. On the issue of military reform in Prussia in this period, see 'Our military institutions' (1819), in the recently published collection of Clausewitz's notes and essays produced between 1803 and 1831, *Historical and Political Writings*.

that all native Prussians were 'bound in duty to the defence of the Fatherland' from the age of twenty.[5] In the long run a more meritocratic ethos prevailed; the national army would come to be drawn from all native-born Prussians, although there were continuing setbacks. After 1819, for instance, the army officer corps again closed its doors to middle-class applicants and the distinction between the regular army and the *Landwehr* widened.

In France, the Napoleonic model of the mass army was abandoned, since the professional officers' desire for a smaller, loyal, politically reliable force with a long-term allegiance to the soldier's trade, and the economic and social desires of the middle class to be exempted from service, coincided. The principle of compulsory service was restored in 1818, but this constituted a recognition of liability to serve rather than actual practice. In fact until 1870 the French army was manned by the unfortunate losers in a ballot of each liable age group. They stayed in the army for a long compulsory period (fixed in 1832 at seven years), often extended voluntarily afterwards, since the army was by then often the only feasible 'trade' for the veteran soldier. The victors in the ballot were simply a hypothetical untrained reserve. Losers, moreover, could pay a 'substitute' to replace them. It was even possible to insure against the risk of being conscripted and needing to pay for a replacement. The relatively affluent were thus easily able to escape the army. Indeed in France the relative poverty of army officers, the lack of interest in developing military training, the want of social cachet associated with the army (the aristocracy sometimes regarding army officers as unreliably 'Napoleonic' in their allegiances) intensified the inadequacies of modernisation. Masked by colonial victories and even to some extent by the Crimea and the Italian campaign, these deficiencies were powerfully revealed in 1870. Just as the victories of Napoleon I had come to be seen as a final 'audit' of the inherent historical weaknesses of his opponents, so the defeat of Napoleon III appeared to reveal some definitive and fatal truth about French moral decline.[6]

The notion that war could set the final stamp upon the moral state or general health of a particular society had been powerfully articulated by both Hegel and Clausewitz. War, both argue, is a forge for the consolidation of national identity, a test of 'virility'. As Clausewitz remarks in *On War*:

> Now in our days there is hardly any other means of educating the spirit of a people in this respect, except by War, and that too

5   See Howard, *The Franco-Prussian War*, p. 11.
6   See *ibid.*, pp. 14–18, upon which I have been drawing extensively in the immediately preceding paragraphs.

under bold Generals. By it alone can that effeminacy of feeling be counteracted, that propensity to seek for the enjoyment of comfort, which cause degeneracy in a people rising in prosperity and immersed in an extremely busy commerce.

A Nation can hope to have a strong position in the political world only if its character and practice in actual War mutually support each other in constant reciprocal action. (p. 262)

Femininity and degeneracy are linked. War channels desire away from 'effeminate' pursuits. Energy flows from the feminine to the masculine principle. On the other hand Clausewitz is generally suspicious of absolute terms and simplistic distinctions. If the objective in war is generally to fight unitedly against a 'foreign' force, the theory of war must nevertheless acknowledge its own inevitable heterogeneity, its inalienable 'foreign' elements:

We must, therefore, decide to construe War as it is to be, and not from pure conception, but by allowing room for everything of a foreign nature which mixes up with it and fastens itself upon it – all the natural inertia and friction of its parts, the whole of the inconsistency, the vagueness and hesitation (or timidity) of the human mind ... (p. 369)

An army fights another. But Clausewitz startlingly reveals that in the conception of war there are never two simple opposing terms, no room for a dogmatic division or binary logic. 'Foreign nature' cannot be excluded from the theory, any more than the reader can simply bypass the metaphors, images and analogies of *On War*. To do so is greatly to impoverish the account. It is the ambiguous and difficult relationship between the maxims of an argument and the narratives, digressions, examples around it which produces the sense of war's complexity. If it is possible to petrify Clausewitz's text into bald propositions, it is also possible to read it novelistically.

It could be argued that *On War* explicitly attempts to uncouple military slaughter from moral questions; that the aim is simply to ignore the issue of the legitimacy or illegitimacy of war, whether as defence or attack, the righting of a wrong, the will of God or the pursuit of good. In his study of the subject, the French social theorist Raymond Aron, for instance, rejects the idea that Clausewitz is a militarist as such. Rather, he is morally indifferent. Clausewitz 'neither condemns nor approves of war, he takes it as a given, primary fact'.[7]

In Clausewitz, it could be said then, the question of right and

7 Aron, *Clausewitz*, p. 101.

wrong is altogether subsumed within a philosophy of pragmatic action, energy, success. But what I suggest here is the 'friction' in the writing of *On War*. Later in this study we will re-encounter such 'friction' in a host of other commentaries and fictions. Yet historians often seek to downplay the contradictions in an attempt to locate a (more highly valued) consistency of message. The historian of Germany and of military thought, Peter Paret, writes 'that despite the unevenness of its execution, *On War* offers an essentially consistent theory of conflict', and that this 'is indicative of the creative power of Clausewitz's method and ideas'.[8] I argue, on the contrary, that the richness of this text – which, as Paret points out, is an ensemble of ideas written over decades – expanded, revised, rewritten[9] – owes a great deal to the tension between its arguments, its suggestive figures, its provisional and uneasy formulations. So the writing itself in a sense exemplifies the impossibility of singular sovereignty over war.

Yet according to Clausewitz war *is* subordinate or at least potentially subordinable to strategic functions and political will. Maverick motives or private interests must be ruled out. The possibility 'that policy may take a false direction, and may promote unfairly the ambitious ends, the private interests, the vanity of rulers, does not concern us here; for, under no circumstances can the Art of War be regarded as its preceptor, and we can only look at policy here as the representative of the interests generally of the whole community' (p. 404). The terms of the exclusion are emphatic. War is 'an instrument of policy' and policy must be understood as the general interest. The sword itself is subordinate to the pen: 'the conduct of War, in its great features, is therefore policy itself, which takes up the sword in place of the pen, but does not on that account cease to think according to its own laws' (p. 410). And yet both the pen and the sword can err, can overrun intentions, become unfamiliar, foreign: 'Just as a person in a language with which he is not conversant sometimes says what he does not intend, so policy, when intending right, may often order things which do not tally with its own views' (pp. 406–7).

In Clausewitz, ostensibly, friction occurs *within* war rather than in the nature of the relation of war to politics. Aron puts it thus: 'After all, in addition to a philosopher of war, Clausewitz could be called a theologian of war. He questions the existence of war no more than the theologian questions the existence of God.'[10] But just

8   Paret, 'The genesis of *On War*', p. 4.
9   See Clausewitz, *Historical and Political Writings*.
10 Aron, *Clausewitz*, p. 15.

as some theologians may well question God, so, in Clausewitz, a more radical interrogation of war is also implied; we glimpse a war machine which threatens the political state with something madder, more disabling and more disruptive than the dominant formulations of *On War* suggest. As Aron acknowledges, Clausewitz's teaching 'will always remain ambiguous' (p. ix).

Clausewitz is difficult to unify in such a way that he becomes unequivocally the counterpart to some other supposedly singular position. It is sometimes argued that the culminating philosophy of Tolstoy's *War and Peace* (1863–9) is quite the antithesis of Clausewitz's theory of war. According to Rapoport, for instance, Tolstoy's philosophy of war is 'the polar opposite of Clausewitz's'.[11] *War and Peace* suggests the primacy of unknown and unknowable forces in war, quite beyond the self-deluding realm of human 'decisions', factors beyond any of the commanders in the Russian campaign of 1812. Where Clausewitz seeks to produce codes and rules for mastering the uncertainties of war, Tolstoy apparently recognises the complete incapacity of either leaders or led to dominate the mystery of fate, the caprice of historical outcomes. But as I have suggested, Clausewitz's text has its own internal conflict. Real wars differ from abstract war just as real machines diverge from 'ideal' machines. There is always an irreducible gap between blueprint and practice; for Clausewitz the question is whether the gap between ideal theory and actuality is amenable to theorising.

According to Clausewitz, Bonaparte ushered in a new age of military power 'based on the strength of the whole nation'. His army 'marched over Europe, smashing everything in pieces so surely and certainly, that where it encountered the old-fashioned Armies the result was not doubtful for a moment'.[12] What one might call the Bonapartist moment in Clausewitz occurs where he imperiously overrides his own narrative to insist, over and over, that whatever happens, politics is in charge. The war of a community, of whole nations and particularly of 'civilised Nations' is by decree always subservient to the political. Where others argue that war is the founding anthropological moment of state self-recognition (Hegel) or disavowal (Proudhon),[13] Clausewitz is adamant that war is born out of the state: 'State policy is the womb in which War is developed, in which its outlines lie hidden in a rudimentary state, like the qualities of living creatures in their germs' (p. 203). The

11 See Rapoport's introduction to Clausewitz's *On War*, p. 16.
12 Clausewitz, *On War*, p. 385.
13 See the discussion in Chapters 4 and 15 (p. 232 ff.).

oft-stated conclusion is that war 'starts from a political condition' and is constrained by a 'political motive' (p. 118). Certainly war opens up a violent and complicated process, but also the interval in which intelligence, control, politics can be exerted – the space, in short, within which war is reined back to the use of a rational will and to the dictate of its mother, the state. Clausewitz decrees it so, but the narrative seems restless and unsatisfied by such extreme and singular directions. War in the real world appears akin to a male sexual drive of very uncertain duration and inconsistent power that needs to be sustained by a 'guiding intelligence':

> it is not an extreme thing which expends itself at one single discharge; it is the operation of powers which do not develop themselves completely in the same manner and in the same measure, but which at one time expand sufficiently to overcome the resistance opposed by inertia or friction, while at another they are too weak to produce an effect; it is therefore, in a certain measure, a pulsation of violent force more or less vehement, consequently making its discharges and exhausting its powers more or less quickly – in other words, conducting more or less quickly to the aim, but always lasting long enough to admit of influence being exerted on it in its course, so as to give it this or that direction, in short, to be subject to the will of a guiding intelligence. (pp. 118–19)

The singular purpose of Clausewitz's own discourse on war threatens to crack, faced by factors which can never be entirely excluded and which may prove disastrously unsettling: 'The whole retains the whole, and as with glass too quickly cooled, a single crack breaks the whole mass' (p. 258). In a successful army or a coherent explanation of war, a certain latitude and flexibility must be tolerated. Yet flexibility can quickly become chaos. History and philosophy continually confront the human subject with new dilemmas; these must be faced to avoid stagnation. But forces can be engendered, fantasies allowed in, armies and weapons unleashed, mechanical, industrial and scientific processes inaugurated, which exceed our grasp. The army general, the scientific inventor or the military theorist is caught up in an interminable quest to hold the internal and external forces of disruption in check, to put paid to the anarchy of new inventions and wayward events. The supposedly instrumental quickly becomes the recalcitrant deviant tormentor. *On War* shares something of the same conception as *Frankenstein, or the Modern Prometheus*, which had appeared a few years earlier. As the creator of the monster, the ironically named Victor, declares:

I considered the being whom I had cast among mankind and endowed with the will and power to effect purposes of horror, such as the deed which he had now done, nearly in the light of my own vampire, my own spirit let loose from the grave and forced to destroy all that was dear to me.[14]

It is moreover around that same period that madness itself is powerfully conceived as an automatism involving precisely the loss of reason.[15] To be too much of an automaton is dangerous, but to give free rein to the imagination and the anarchy of feeling is also risky. Clausewitz's text dramatises the problem of domination which also constitutes its subject. 'Friction', 'chance', 'the unpredictable' keep threatening to slide out of control in the Clausewitzean treatise itself, challenging the procedures, insights and pedagogic examples which the author so carefully delineates for the benefit of war-makers. The text is both 'on' war and itself enmeshed in a conflict of ideas and images. Words are recognised here as the site of a problem. Clausewitz acknowledges the unwieldy and defiant propensities of even the most rigid language, warning that technical terms may threaten a rearguard assault, provoking chaos in the system:

Much greater is the evil which lies in the pompous retinue of technical terms – scientific expressions and metaphors, which these systems carry in their train, and which like a rabble – like the baggage of an Army broken away from its Chief – hang about in all directions. (p. 229)

Clausewitz constantly shifts the optical perspective, moving back and forth between the Enlightenment and its romantic shadows, between an age of rationalism and its subsequent critique, sliding away from the vantage point of 'the eyes of reason' (p. 291) to which the text is nevertheless committed. The machine constitutes one of the most deeply ambiguous and troubling central analogies of his story of war.

Above all, it is friction which distinguishes real war from war on paper. The military machine may *appear* to be a single thing, but since it is composed entirely of individuals who all produce friction,

14 Mary Shelley, *Frankenstein*, ch. 7, p. 117.
15 As Michel Foucault writes with regard to the great French psychiatrist of the Revolutionary period, Pinel: 'The truth of madness lies in an unchained automatism; the more an act will be empty of reason, the more chance of its being born in the determinism of madness; the truth of madness in man being the truth of that which is without reason' (*Folie et déraison*, p. 622).

the commander of a battalion is faced with vast problems: 'This enormous friction, which is not concentrated, as in mechanics, at a few points, is therrefore everywhere brought into contact with chance, and thus incidents take place upon which it was impossible to calculate, their chief origin being chance' (p. 165). War can never be converted into geometry. 'The conduct of war resembles the working of an intricate machine with tremendous friction, so that combinations which are easily planned on paper can be executed only with great effort.'[16] It is precisely because war is in the realm of the human and the moral that it cannot be an abstract mathematics.

From the eighteenth to the early nineteenth centuries we find in war theory a continuing interplay of the elaboration of new dynamic systems and the critique of petrified mathematical models. As Aron pointed out and as the military historian Azar Gat interestingly elaborates, Clausewitz finds in contemporary art theory a highly suggestive model for his military convictions.[17] War is the scene of creativity, genius and energy, as well as logic, planning and science. Clausewitz rejects dead abstractions, insisting on the importance of 'subjective' forces: life, character, spirit, genius. In his Letter to Fichte (1809) Clausewitz criticises 'the tendency, particularly in the eighteenth century, [to] turn the whole into an artificial machine in which psychology is subordinated to mechanical forces that operate only on the surface'.[18] Soldiers cannot be treated 'like simple machines'. The spirit of war consists 'in mobilizing the energies of every soldier to the greatest possible extent and in infusing him with the warlike feelings, so that the fire of war spreads to every component of the army instead of leaving numerous dead coals in the mass' (*ibid.*).

Previous attempts to establish maxims for the conduct of war always foundered, Clausewitz declares in *On War*, because they were produced on the basis of quite arbitrary calculations, a whole array of geometrical assumptions which can be dismissed as 'utterly useless' in practice (p. 183):

> We see, therefore, how, from the commencement, the absolute, the mathematical as it is called, nowhere finds any sure basis in the calculations in the Art of War; and that from the outset there is a play of possibilities, probabilities, good and bad luck, which

16 Clausewitz, *Principles of War*, p. 50.
17 See Aron, *Clausewitz*; Gat, *Origins of Modern Military Thought*, pp. 176–7. Clausewitz made statements which whilst rejecting mysticism insisted on 'the need to return from the tendency to rationalize to the neglected wealth of feeling and fantasy' (see Gat, *Origins*, p. 184).
18 Clausewitz, 'Letter to Fichte', in *Historical and Political Writings*, p. 282.

spreads about with all the coarse and fine threads of its web, and makes War of all branches of human activity the most like a gambling game. (p. 117)

In another analogy, war belongs more to the field of commerce – the comparison which so interested Engels.[19] Clausewitz proposes a science of war which will not simply be a closed system, which recognises the danger of abstraction and formalism, and which insists that people are not predictable machines. Too inflexible a set of tactics 'leads to an Army made like an automaton by its rigid formations and orders of battle, which, movable only by the word of command, is intended to unwind its activities like a piece of clockwork'.[20] Bonaparte was quite right, Clausewitz insists, to say that war raises problems for algebra before which even a Newton might stand aghast (*ibid.*, p. 375). The modern theorist must resist formalism, abstraction, absolute mathematical certainties: 'War must always set itself free from the strict law of logical necessity, and seek aid from the calculation of probabilities' (p. 125). Indeed chance is an intruder which destabilises the field of war to a greater extent than any other sphere of politics.[21]

Just as the biological sciences were emerging against the grain of eighteenth-century taxonomy, insisting on growth, history, development and evolution, so war theory was to be reconceived in Clausewitz as an analysis of historical-evolutionary change. He indicated that 'finished theoretical constructions' are no longer central to science; the 'scientific form' lay in the endeavour to explore historically the enduring and shifting aspects of military phenomena.[22]

Clausewitz has many analogies for war; each has its specific resonances and takes war in a different imaginative direction. If on the one hand war is to be thought of as a kind of machine, on the other it is to be conceived as a contest between two autonomous subjects, 'nothing but a duel on an extensive scale'. A duel, or then

19 See Semmel's discussion of Engels on war in his *Marxism and the Science of War*, and my discussion in Chapter 5, below.
    For the tension between the ideas of Clausewitz and the rivalrous French military theorist Antoine-Henri de Jomini (1779–1869), in relation to questions of predictable law and mathematics, see Gat, *Origins*, p. 130.
20 *On War*, p. 181.
21 'War is the province of chance. In no sphere of human activity is such a margin to be left for this intruder, because none is so much in constant contact with him on all sides. He increases the uncertainty of every circumstance, and deranges the course of events' (p. 140).
22 See *ibid.*, p. 91.

again a wrestling match: 'If we would conceive as a unit the countless number of duels which make up a War, we shall do so best by
supposing to ourselves two wrestlers' (p. 101). The struggle, once
under way, should rightly play itself out to the bitter end.[23] The
two fighters could also be seen as two energy systems. Repeatedly
Clausewitz draws on images of force and resistance to characterise
the nature of armies and leaders, above all the Napoleonic army.[24]

To view war as a static system is always to court disaster,
Clausewitz explains. Against the petrified code, Clausewitz insists
on the disruptive force of energy – the intensity and power previously encapsulated by Frederick the Great and now epitomised
above all by the challenge of the French Revolution and its emperor:

> Has not then the French Revolution fallen upon us in the midst of
> the fancied security of our old system of War, and driven us from
> Chalons to Moscow? And did not Frederick the Great in like
> manner surprise the Austrians reposing in their ancient habits of
> War, and make their monarchy tremble? Woe to the cabinet
> which, with a shilly-shally policy, and a routine-ridden military
> system, meets with an adversary who, like the rude element,
> knows no other law than that of his intrinsic force.[25]

The *levée en masse*, the innovation of 1793, which began as a
Romantic idea and a response to an immediate crisis, mobilising
vast numbers of peasants, often armed only with pitchforks and
hunting knives, was to evolve, as Simon Schama puts it, 'into a
professionally organised and highly disciplined arm of the state'.[26]

23 'A suspension in the act of Warfare, strictly speaking, is in contradiction with
    the nature of the thing; because two Armies, being two incompatible elements,
    should destroy one another unremittingly, just as fire and water can never put
    themselves in equilibrium, but act and react upon one another, until one quite
    disappears' (p. 291). Where Clausewitz insisted that war should be totalised,
    others were asking, sardonically, why it should not be minimised. *Cf.* the
    English social critic Thomas Carlyle's reference to Smollett in *Sartor Resartus*:
    'In that fiction of the English [*sic*] Smollett, it is true, the final Cessation of
    War is perhaps prophetically shadowed forth; where the two Natural
    Enemies, in person, take each a Tobacco-pipe, filled with Brimstone; light the
    same, and smoke in one another's faces, till the weaker gives in: but from such
    predicted Peace Era, what blood-filled trenches, and contentious centuries,
    may still divide us!' (*Sartor Resartus*, p. 120). Tolstoy raises the same question
    in *Tales of Sevastopol. The Cossacks*, p. 40.
24 'In these operations, and especially in the campaigns of Buonaparte, the
    conduct of War attained to that unlimited degree of energy which we have
    represented as the natural law of the element' (p. 291).
25 *Ibid.*, p. 294.
26 Schama, *Citizens*, p. 760. *Cf.* more generally pp. 760–5, on the
    professionalising of the recruits, the restoration of discipline and the vast

It was Napoleon above all who triumphantly galvanised the new mass military machine.[27]

War is an act of violence intended to compel an opponent to fulfil our will. Clausewitz emphasises its reciprocity and its duration – it takes two, it is a duel and its outcome is not instantaneous. 'War is always the shock of two hostile bodies in collision, not the action of a living power upon an inanimate mass' (p. 104). The duration of war is a crucial feature, for it is in the interim of the conflict that all the forces converge in a form that is never entirely predictable. Unexpected courage, genius, folly and sheer chance collide with other factors: planning, numbers, strategy, tactics and strength.[28] He rejects the idea that war might become excessive, freewheeling of its own accord; yet such a prospect returns as a nagging hypothesis, an old supposition to be discarded more than once:

> Now if it was a perfect, unrestrained, and absolute expression of force, as we had to deduce it from its mere conception, then the moment it is called forth by policy it would step into the place of policy, and as something quite independent of it would set it aside, and only follow its own laws, just as a mine at the moment of explosion cannot be guided into any other direction than that which has been given to it by preparatory arrangements. This is how the thing has really been viewed hitherto, whenever a want of harmony between policy and the conduct of a War has led to theoretical distinctions of the kind. But it is not so, and the idea is radically false. (p. 118)

The hierarchy of control, the relationship of subject (political reason) and instrument (war) must be retained at all costs: 'in the modifications of reality [war] deviates sometimes more, sometimes less, from its strict original conception, fluctuating backwards and forwards, yet always remaining under that strict conception as under a supreme law' (p. 137). As though appalled by glimpses of the uncontrollable, Clausewitz falls back on the concept of simple

---

expansion of military service and the arms industry. According to one estimate, by the spring of 1794 6,000 workshops were apparently busy making gunpowder (*ibid.*, p. 765).

27 The Napoleonic legend begins with Napoleon himself and persists across the century. As he declared: 'I had the gift of electrifying men' (quoted in Geyl, *Napoleon: For and Against*, p. 146). As Geyl shows, from Napoleon to Taine, Sorel and Barrès, the Emperor was viewed as the apostle of energy.

28 See Clausewitz, *On War*, Book 4, ch. 6, 'Duration of the combat'. For instance, pp. 318–19: 'For the conqueror the combat can never be finished too quickly, for the vanquished it can never last too long. . . . The duration of a combat is necessarily bound up with its essential relations.'

commerce. Like the idealised calculating subject of classical political economy, the war commander must weigh the odds, rationally and reliably. War is a commodity to be bought and sold only when the price is appropriate, the outlay not exorbitant:

> As War is no blind act of passion, but is dominated by the political object, therefore the value of the object determines the measure of the sacrifices by which it is to be purchased. This will be the case, not only as regards extent, but also as regards duration. (p. 125)

Questions of friction, illness, madness, morals, fear and anarchy continuously need to be mastered by this war theorist, converted back into a manageable currency which enables decision-making. He presides over and marshalls his thoughts, like a general seeking to retain control over potentially wayward troops. War is chameleon-like in character, we are told, each 'case' appearing distinct. Even so, it is always structured, as Clausewitz puts it, around 'a wonderful trinity' of elements: first, the original violence of its impetus and components – the power and capacity of relative hatred and animosity (a subject which receives fairly short shrift); second, the play of probabilities and chance (which, apparently, 'make it a free activity of the soul'); third, its subservience to politics (war has 'the subordinate nature of a political instrument, by which it belongs purely to the reason', p. 121). Clausewitz's notion of 'friction' has to be held in check, never allowed to double back on the stated certainty that the human subject is, in its potentiality if not always in its practice, above all else a unified reasoning being.

The subjective experience of battle, with all its attendant dangers, is an element of 'the friction' to be taken into account. Clausewitz invites his readers to 'accompany the novice to the battlefield', asking that we imagine the thunder of the cannon becoming plainer and plainer, the howling of shot, the striking of cannon balls and the sound of explosions all about. We must envisage how the death and maiming of those around the novice provokes his compassion, how the grape rattling on roofs arouses a shudder. Above all the novice will recognise that 'the light of reason does not move here in the same medium, that it is not refracted in the same manner as in speculative contemplation' (pp. 159–60). But the question of whether war can derange reason altogether – drive the participants mad – is not addressed; simply the fact that 'an ordinary character [faced by the battle] never attains to complete coolness and the natural elasticity of mind' (p. 160). Here and there Clausewitz returns to the question of human frailty, caprice and waywardness.

These are issues to be subsumed in the ultimately rational strategy of war:

Action in War, therefore, like that of a clock which is wound up, should go on running down in regular motion. But wild as is the nature of War it still wears the chains of human weakness, and the contradiction we see here, viz. that man seeks and creates dangers which he fears at the same time will astonish no one. (p. 291)

Any sphere in which 'the moral activities' and 'bodily phenomena' are at work subverts the possibility of abstract rules and formulae. Hence the science of medicine faces constant change and uncertainty between one case and another. The doctor of the body can never rely simply on mathematical science: 'but how much more difficult is the case if a moral effect is added, and how much higher must we place the physician of the mind?' (p. 185)

4 Proudhon's *War and Peace*

In the previous chapter I explored ambiguities of image and viewpoint in Clausewitz's *On War*. I now turn to a very different and less well-known account of war by the French socialist and anarchist theoretician, Proudhon, in which there is also a powerful friction between maxim and language, manifest intention and unfolding argument. In *War and Peace* (1861), Proudhon insists upon war's fundamental primacy in politics; indeed the political state is a war manifestation. On display here is not so much 'the poverty of philosophy' (as Marx put it in deriding his rival's *Philosophy of Poverty*), but rather its extreme provocation, in which the writer's own supposedly pacific aims are violently undone by the seduction of war. It is as though Proudhon insists that we look not only at war's social effects, but at the havoc it plays with his own argument. Where Clausewitz theorised the nature of war in relation to Napoleon I, Proudhon's treatise was written in the more burlesque atmosphere of Louis Napoleon's Second Empire. Yet it was also powerfully affected by the intellectual shadow of Hegel.[1]

The Eighteenth Brumaire of Louis Napoleon begins with Marx's great 'riposte' to Hegel:

> Hegel remarks somewhere that all facts and personages of great importance in world history occur, as it were, twice. He forgot to add: the first time as tragedy, the second as farce.

As though commensurate with that contemporary sense of grotesque pantomime, Proudhon's reflections on war are not without their bleak comedy. Writing supposedly as a critic of war justifications, Proudhon enters into a vast 'devil's advocacy' of state violence. This takes on such elaborate and memorable proportions as to crowd out and relegate the critique. Something of the war enthusiasm under investigation runs out of control, infecting the text, to the chagrin of some of Proudhon's later commentators. His sympathetic biographer George Woodcock acknowledges that *War and Peace* involves 'complex and at times, perverse arguments', although Proudhon apparently reassures him in the end about his

---

1 On Proudhon's relation to Hegel, see Hoffman, *Revolutionary Justice*, pp. 89–94. Marx claimed to have 'infected' Proudhon with Hegel at their meeting in Paris in 1844; see *ibid.*, p. 89.

good intentions: 'At times the passion for stating both sides of an argument leads Proudhon into talking like a frenzied devotee of militarism, but even here, as he states the affirmative side of war, there are twists of argument that reassure the perceptive reader.'[2]

As the bloody days of 1848 continued towards their catastrophic climax, Proudhon had perceived that the conflict dramatically symbolised the endemic war beneath society and the persistent treachery of representative government. He wrote in his diary on 25 June 1848: 'The Terror reigns in the capital, not a Terror like that of '93, but the Terror of the civil and *social* war.'[3] Proudhon recorded the horror of the fratricidal violence manifest in the capital. Intoxicated soldiers were 'unleashed like bulldogs on the insurgents.... Horror, horror!'[4] When the last burst of fighting took place around the Bastille on 26 June, Proudhon was there as a witness, in the hope, as he told the subsequent investigating commission, 'of leading the strayed sheep back'. Proudhon's temerity in rushing to the streets as soon as the barricades went up placed him in danger under the new reaction. A strong attempt was made to implicate him in the responsibility for the uprising on the grounds that he had been seen within the barricades before the soldiers entered. Questioned about his purpose at the Bastille, Proudhon grandly declared that he had been listening to the sublime horror of the cannonade.[5] Whatever sublime sounds Proudhon may have heard in 1848, he was increasingly dismayed by the people's apparent acquiescence in, or even enthusiasm for, Louis Napoleon's plebiscitary despotism even though he himself had disclosed some enthusiasm for the return of Bonapartism.[6] He was appalled by, amongst other things, public support for the emperor's military adventurism in the Crimea and Italy, with its pathetic echoes of Napoleon I. He decried current moral decadence and the squander-

---

2 Woodcock, *Proudhon*, pp. 233–4. *Cf.* '*War and Peace* was a book made to arouse controversy, and the apparent contradictoriness of its theme was given an almost grotesque emphasis by the way in which it was presented.' (*Ibid.*, p. 234).

3 Proudhon, *Carnets*, vol. III, p. 66.

4 *Ibid.*, p. 67.

5 See Woodcock, *Proudhon*, p. 131.

6 This stemmed from his increasing guilt about and hostility towards the course of the 1848 revolution, the view that all representative government was a fraud. See Cogniot, *Proudhon et la démagogie bonapartiste*; and Hoffman, *Revolutionary Justice*, pp. 198–99, 206. Bonaparte's apparent advantage lay in his intermediate position between the two extremes of revolution and reaction. For Proudhon's attacks on Louis Bonaparte, see Hoffman, *Revolutionary Justice*, p. 136.

ing of resources, clerical hypocrisy, social misery and military brutality.[7]

War both constituted and undermined government authority; in other words the state was founded on a force which destabilised it. In *War and Peace* Proudhon undertakes a general interrogation of state legitimacy. By insisting that war is always legitimate since all states are founded on war, he pulls the rug simultaneously from under those who oppose war and those who legitimise the state. War is shown to precede and underpin governments, and thus to haunt their supposedly pacific objectives. Command is ultimately based on that fact of power and conquest; yet governments habitually veil and disavow their own bellicose origins, compounding thereby an underlying insecurity in their very structure, as though war is the unmentionable skeleton in the cupboard of the social formation.

Yet Proudhon is apparently inconsistent. War is viewed both as structurally indispensable and as potentially transient. It is indispensable, however, only in the current social order. He thus holds out the possibility of finally overcoming war in the future. He shifts its motivation from desire to need; so it is the constant pressure of poverty and the absence of subsistence that forces society into battle. Nevertheless Proudhon, like Hegel, rejects Kant's 'perpetual peace' as unrealistic. War, he suggests, cannot be theorised abstractly; it demolishes the very possibility or at least utility of disembodied philosophical categories and heralds the triumph of practical over pure reason: 'Perpetual peace is IMPRACTICABLE.'

The specific gesture of rejecting Kant's 'perpetual peace' as unrealistic was to be an increasingly familiar trope; new credentials for that rejection would emerge after Hegel and Proudhon in a later nineteenth-century era which so often combined idealism and mysticism with an increasingly pervasive evolutionary naturalism. Consider Field Marshal von Moltke's comment, reported with trepidation in 1881 in *The Times*: 'Perpetual peace is a dream, and it is not even a beautiful dream.'[8] The belief in the evolutionary and

---

7   See Woodcock, *Proudhon*, pp. 195–6; Hoffman, *Revolutionary Justice*, p. 218.
8   See Moltke's letter on war and the newspaper's editorial comment, *The Times* (1 February 1881), pp. 3, 9–10.
    Moltke was referring to a manual on the laws of war sent to him by the Institut de Droit International which had recently held a meeting in Oxford. The Field Marshal had replied in a letter (translated, printed and discussed in *The Times*) which, whilst regretting the evils of war, rejected the legists' view of it as an aberration. It was, he insisted, an element of the order of the world

psychological necessity of war had been forcefully set out by Hegel and it would be re-expressed and reconsidered in numerous subsequent commentaries through the nineteenth and early twentieth centuries, not least by Freud.[9]

But to return to Proudhon: since thought itself involves conflict, war cannot be abolished from society by some imperious gesture of thought itself. War is everywhere. Proudhon recognises a progression in society from armed struggle to labour and industrial struggle, but insists that it must not be forgotten that those higher stages are themselves founded on bloodshed.[10]

War may change its form but its essential structure and causation remain fixed. Conquest is inevitable in a world of endemic poverty.[11] Proudhon complains that the conventional philosophical discussions of war hear only its 'noise' and see only its stock images. They are unable to grasp its essence: 'Nobody has yet sought to seize on war's innermost thought, its reason, its consciousness, to be precise, its high morality' (p. 24).

War explodes pure reason: it continually stresses the gap between blueprints and actual experience. Why, asks Proudhon, is there always so stark a divide between the ideal fiction of the conduct of war and its reality? How can one explain the perennial contrast between elaborate military codes of honour and the systematic brutality of armies in practice? At stake in military violence and atrocity, he insists, is not simply the issue of the odd maverick soldier committing evil deeds, or rampaging army groups out of

---

ordained by God. The soldier was the highest example of human generosity. Indeed if Moltke appreciated the Institute's efforts, *The Times* noted caustically, this was because it might humanise the suffering of war for the soldier rather than the civilian. In war, 'the noblest virtues of mankind are developed; courage and the abnegation of self, faithfulness to duty, and the spirit of sacrifice: the soldier gives his life. Without war the world would stagnate, and lose itself in materialism.' Nevertheless, Moltke kept insisting that war was not a tea-party. An invading army could not be expected to negotiate food and accommodation on the 'free market' as might be expected of tourists. It was a mistake to seek to place curbs and limits on war. 'The greatest kindness in war is to bring it to a speedy conclusion.' Hence it was right to attack all the resources of the enemy's government, its finances, railways, stores and even its prestige. *The Times* contended that it was in the self-interest of nations to fight according to certain codes, just as it was in duelling. See my discussion of Treitschke and von Bernhardi in Chapter 8.

9 See Chapter 15.

10 See Proudhon, *La Guerre et la paix*, p. 498.

11 'To conquer, conquer again, and always conquer; since the dawn of societies, and by virtue of the pauperism which is endemic to them, that has been the tendency of States.' See *ibid.*, p. 401.

control, but a structural feature of civilisation itself. Violence and atrocity underpin it.

For all its illusions, Proudhon considers war 'the most profound and sublime phenomenon in our moral life' (p. 489). He suggests, like Clausewitz, that war is shaped by a 'trinity' of factors: for Proudhon these are force, antagonism and justice (p. 469). War touches on something incorruptible in the human being and something of the essential dialectic that structures history, life, the soul, legislation, politics, the state, the fatherland, the social hierarchy, rights, poetry, theology, everything (p. 71).

Despite Proudhon's pronouncement that war may be transient, it turns out that really it is inseparable from the nature of human identity and society. It is constitutional in the deepest sense, in effect the primary human right and human relation (p. 496). Accordingly, the claim that war is legally unconstitutional is merely absurd. To deny a 'right' to war, he argues, is to represent existing governments as a mere random lottery, an arbitrary collection of entities meaningless in itself. The question of war continually leads back to the question of the origin of each state. Without positing a right to war, the state (always and inevitably founded on violence) can have no moral authority.

War cannot be externalised and alienated from humanity, since human identity itself – the very production of human being – is founded on war. Modern philosophy characteristically fails to understand that war is 'an internal rather than an external phenomenon.... [It] is divine, which is to say primordial, essential to life, to the very production of man and society' (p. 23). This insight violently collides with Proudhon's suggestion that war will disappear with the disappearance of need.

Proudhon argues that peace can be defined only in relation to war, indeed as its absence. The terms, which 'the vulgar' deem mutually exclusive opposites, are in effect meaningful only reciprocally. They are the two inseparable sides of the same coin (see pp. 63–4). There can be no final end to war without the dissolution of society, since Proudhon has argued that community itself is not only structured upon war but also revitalised in conflict. A society with no will to fight has lost its soul; conversely the one that keeps this 'righteous flame of war alive in its heart' is the force of the future (p. 91). War cannot be understood, he insists, as anachronistic, like say piracy or brigandage; on the contrary, at its best war constitutes the confrontation of two vanguards, two elites, in a highly formalised encounter.

Proudhon's text operates a continual strategy of provocation. To reject his argument, he apparently implies, would amount to anarchy; for to reject the idea that war is constitutional is to deny the legitimacy of any society founded on armed force – which is to say, in Proudhon's terms, all organised societies. Without a right to war, 'chance is king' and the social system rests on a mere 'figure of rhetoric' (p. 96). To oppose Proudhon's argument is apparently to acknowledge the illegitimacy of existing structures of government; to agree with him is to recognise the right of the 'slave' to take the 'master' by force and install a new system of power.

What I have tried to suggest is that to read the war discussions of Clausewitz or Proudhon is to read a complex set of claims and assumptions about power and anarchy, about the achievement of state identity through the location of and victory over the enemy, about freewheeling processes of disorder, about the forces that can be marshalled for and against disintegration. This literature, moreover, is caught up in a dilemma about method – about how to write adequately of war, how to think its terms, how to master its language. These languages of war are never reducible to some singular 'position'; they display the clash of ideas and forces with which they also deal. Clausewitz insists that war is an instrument of policy; indeed war is policy, which merely 'takes up the sword in place of the pen, but does not on that account cease to think according to its own laws' (p. 410). Nevertheless another possibility intrudes: the pen disturbingly caught up in the 'friction' it describes.

5 Engels and the Devouring War of the Future

While Proudhon wrote his *War and Peace*, Engels was enjoying Clausewitz. As he wrote in a well-known letter to Marx at the beginning of 1858: 'Among other things, I am now reading Clausewitz *On War*. A strange way of philosophising but very good on his subject. On the question whether war should be called an art or a science, the answer given is that war is most like trade.'[1]

War is like trade: it is like an exchange of goods; it has a mercantile ethos, involving the weighing up of prices and returns. To put it another way, war is about trade and not about industry, because in the model of trade it is possible to imagine two autonomous and sovereign reasoning beings making their calculations, using war as their money. In fact, the analogy of war to trade is only one among many to be found in Clausewitz's *On War*. The message that 'war is trade' would be a very partial characterisation of the story *On War* tells.

Clausewitz wrote in the early part of the nineteenth century; he was reread after 1870 as the apostle of the 'modern technological war' of the 1860s and beyond. His cool acknowledgement of the expansive and eventually totalising demands that the war machine may place on the state that utilises it makes Clausewitz seem 'modern'. He spoke after all of 'modern absolute war in its destroying energy'.[2]

In one conception the changes taking place across the century

---

1 Marx and Engels, *Correspondence*, p. 100. A few months earlier Marx had emphasised the importance of the army: 'The history of the *army* brings out more clearly than anything else the correctness of our conception of the connection between the productive forces and social relations' (pp. 98–9). For a recent corrective to the view that Marx and Engels in any simple sense derived their views on war from Clausewitz, see Gat, 'Clausewitz and the Marxists'. Gat argues that the apparent affinity stems less from the direct influence of Clausewitz than from the fact that all these writers shared the same German historicist tradition.

2 See Clausewitz, *On War* (Penguin, 1986), p. 373: 'At the time of the Silesian War in the eighteenth century, War was still a mere Cabinet affair, in which the people only took part as a blind instrument; at the beginning of the nineteenth century the people on each side weighed in the scale.... It is just those very campaigns of 1805, 1806, 1809, and following ones, which have made it easier for us to form a conception of modern absolute War in its destroying energy.' See also Howard, 'The Influence of Clausewitz'.

in the capacity for 'destroying energy' could be characterised as rationalisation. War, it could be argued, was increasingly being pushed beneath the sway of programmed and efficient technology, managerial planning, factory-produced machines with ever greater devastating power. But what we can also trace in the later nineteenth-century period, sometimes in the margins of the very texts which celebrated modernity as the field of technological 'progress', is another conception: the machine as agent of an open-ended process of destruction and disaggregation beyond the question of any political will or limit. The point here is not that we are simply offered further versions of a Tolstoyan vision of the irrelevance of human command in the face of unknowable fate in war, but that we encounter a deranged machinery. This is not instrumental war, not war as trade, but the delirium of technology – the machine as protean, destructive and monstrous. Clausewitz is so often on the brink of this image. The discursive shift towards such a conception, such an image, in the later nineteenth century is central to my argument. This was not exclusively the retelling of *Frankenstein*,[3] the romantic drama of the author/scientific explorer transformed into 'the slave of [his] own creature'.[4] Nor was it just one further twist to a Classical drama in which violence, once unleashed, knows no limits. We can locate new elements, new stories and horrors in relation to technology, destruction and national identity, glimpsed possibilities of the irreducible ambiguity of the modern war machine. It is precisely the historical complexity of this presentation of war as deranged machine which we need to reconstruct.

In practical terms, it has been argued elsewhere, the military implications of industrialisation had not been substantially explored or exploited in the first half of the nineteenth century. In Britain, for instance, the combination of Treasury reluctance and War Ministry conservatism tended to stifle any ambitious scientific developments until the Crimean War brought home not only that major wars were still possible, but also that those wars would be decided on the basis of new technology. Throughout Europe military writing on the ramifications of the industrial revolution increased in the second half of the century. Between 1815 and 1848 European armies had been mainly concerned with and geared to domestic policing or at least counter-insurgency operations. Armies on the scale of the Napoleonic period no longer seemed to be necessary. For such

3 Although there were in fact numerous retellings of *Frankenstein* across the century: see Baldick, *In Frankenstein's Shadow*.
4 Shelley, *Frankenstein*, p. 239.

internal purposes, the dynastic loyalty of the troops was crucial. This seemed to favour the maintenance of relatively small forces of long-serving regular troops commanded by exclusively aristocratic officer corps. In short, the first half of the nineteenth century saw the re-emergence of conservative military traditions and practices that partially went against the grain of the Revolutionary wars of masses, of the whole logic of universal conscription.[5]

Already in the 1830s, however, German writers had been quick to see the militarily transformative properties of railways, particularly for a weak German Confederation potentially exposed to a resurgent and expansionist France. Writers like Friedrich List argued that the railway would help unify the German economy and would enable the army to gather itself quickly at the specific site of tension. The railway would thus realise the Napoleonic maxim of ensuring the concentration of overwhelming force at the decisive point. 'Every new development of railways,' von Moltke was to remark, 'is a military advantage; and for the national defence a few million on the completion of our railways is far more profitably employed than on our new fortresses.'[6]

In place of exhausted troops who had marched long distances with vast burdens, the railway often enabled troops to arrive at their destination in good condition to be fed, clothed and rearmed by rapid and more reliable supply lines. In practice, of course, 'friction' could and did occur in railway supply, as everywhere else in military organisation, with clogged routes, inappropriate provisions, mistaken destinations, poor coordination and so forth. Instead of a Germany vulnerable to attack by the concerted action of her neighbours, however, it became possible to envisage a Germany capable of exploiting its continental centrality to mobilise troops quickly at any point on the frontier. Communications were facilitated within the military system itself and between the battlefield and the wider society, in both Europe and America. Newspaper reporters could move back and forth relatively easily via the railway and stories could be sent via the electric telegraph.

The battlefield's modern dispersal and increasing technical com-

---

5 As Michael Howard writes: 'Everywhere armies languished in unpopular and impoverished isolation, and few thinkers in a Europe increasingly wealthy, materialist and optimistic saw in this any cause for regret. Even the restored and chastened French monarchy reconstructed its military system on a basis as alien to the Napoleonic as possible; and the state, curiously enough, which preserved most of the apparatus of Revolutionary military organisation was that paragon of all conservative regimes: the Hohenzollern monarchy of Prussia' (*The Franco-Prussian War*, p. 10).

6 Quoted in *ibid.*, p. 2.

plexity made it ever more difficult for its course to be controlled from a single vantage point or its participants orchestrated by a lone commanding will. Specialist functions had to be delegated more widely. The notion of some single autonomous intelligence or genius in control of the whole operation, always questionably romantic and idealist to say the least, became ever more far-fetched. The army, in short, was increasingly grasped as a complex body with an elaborate division of labour. In principle, the relative autonomy of specialists, subordinate officers and their troops could and indeed needed to be set within the dictates of an overarching policy, but this was a difficult business. As Michael Howard writes of the Prussian case:

> On the great fields of manoeuvre, moreover, there was little opportunity for direct control. An army headquarters might be several days' march from its forward troops when they engaged; and if the will of the commander was to make itself felt at all, it could only be through reflexes which he had already inculcated in his subordinates through previous training; so that, even when deprived of his guidance, they should react to unexpected situations as he would wish.... Thus the Prussian General Staff acted as a nervous system animating the lumbering body of the army, making possible that articulation and flexibility which alone rendered it an effective military force; and without which the French armies, huddled together in masses without the technical ability to disperse, found numbers a source not of strength but of fatal weakness.[7]

French and Austrian commentators also soon comprehended some of the ramifications of the railway and the challenge it posed to traditional strategies and balances. In the struggle for Italian unification, both the French and the Hapsburg Empires made use of railways to send troops into Italy, reducing various journey times, according to Howard, to a quarter of their former length.[8] Firepower was also transformed. In place of laborious, slow-loading muskets whose accuracy at a distance even of 50 yards was far from sure, by 1870 the large European powers possessed rifles accurate

---

7　*Ibid.*, p. 24.
8　See *ibid.*, p. 3. On the insuperable political, economic and social difficulties that Prussia's adversaries, especially Second Empire France, faced in matching or even fully grasping her military reforms in the 1860s, see *ibid.*, pp. 29–30, 39. Thus in the late 1860s the French army *did* prepare intensely for war with Prussia, but a bygone form of war still associated with the Crimea, the Italian War and the Mexican adventure.

up to at least 500 yards with rifled guns which could be used at ranges of 2–3,000 yards.

But whatever the conceptual and industrial developments apparent from the 1830s onwards, the organisational transformation of Prussia's hitherto relatively weak military power and performance (as compared to France) was remarkably sudden and swift in the 1860s. It followed the series of reforms proposed by Albrecht von Roon to the new Regent and former military man, Prince William of Prussia, who had taken over from the incapacitated Frederick William IV in 1858. Obstacles to the establishment of a new, well-trained, short-service conscript army were swept aside. The unenthusiastic and obstructive General Edouard von Bonin was quickly replaced by the ultra-reactionary Roon himself as Minister of War in 1859. The *Landwehr* was reorganised, the regular army was increased in size; training and terms of service were improved. Despite opposition, the enduring conflict between Crown and Assembly (dissolved in September 1863, a year after Otto von Bismarck became Minister-President), the increased military budgets were pushed through. War with Denmark in 1864 and Austria in 1866 led the restored Assembly to approve retrospectively the army reforms and increased costs.[9]

I have sketched out some of these developments in army organisation and machinery, particularly the adaptation of the railway to military logistics, in order to contextualise the transformation taking place in the very conceptualisation of war. This may run the risk, however, of erroneously suggesting that the 'material' changes in 'hardware' were somehow prior to or outside of the changes in perception with which, in fact, they were inextricably bound up. Writing on war is part of the material culture and economy it represents, not some subsequent straightforward effect. The 'war of words' would later be an active and concrete part of the history of the First World War itself, which is not to say that words alone produce wars. Language shapes and is shaped by social, economic, political and technological change. Representation and the pressure of symbolisation in general is never a straightforward 'reflection' of some other material 'bedrock', some supposedly non-discursive economic stratum, for instance, that can be privileged as determining 'in the final instance'.

Marx shared with Cobden the view that militarism as such was

9 For a more detailed discussion of the army reforms, in part copied or adapted across the new North German Confederation, see Howard, *The Franco-Prussian War*, pp. 19–22. By 1870 the North German Confederation was reckoned to have more than a million men in arms; see *ibid.*, p. 22.

fundamentally anachronistic and atavistic. But in the very notion of a machine that alienates the individual, which he and Engels theorised, there was also the basis for a rather different vision of the army and military power to emerge, a vision relevant to the world of a revamped Prussia. And indeed Engels would come to explore, where Marx had not, a modern technological future of war and trenches, stagnation and the death of millions, beyond the question of 'Clausewitzean' political purpose, beyond the reflection of any 'interest'.

Marx and Engels had already been insisting in the 1840s and 1850s that the dehumanisation of the worker under modern factory conditions constituted not only alienation but a new order of regimentation, a virtual militarisation of economic relationships:

> Modern industry has converted the little workshop of the patri-archal master into the great factory of the industrial capitalist. Masses of labourers, crowded into the factory, are organized like soldiers. As privates of the industrial army they are placed under the command of a perfect hierarchy of officers and sergeants. Not only are they slaves of the bourgeois class, and of the bourgeois State; they are daily and hourly enslaved by the machine, by the overlooker, and, above all, by the individual bourgeois manu-facturer himself.[10]

The 'various metamorphoses' of modern society and economy cul-minate in the '*automatic system of machinery*'. Labour is mechanised, and the machine is gradually transformed into a system. The machine is not a tool manipulated by the worker but vice versa. The process is 'set in motion by an automaton, a motive power that moves itself; this automaton consisting of numerous mechanical and intellectual organs, so that the workers themselves are cast merely as its conscious linkages'.[11]

With the artisan's tool it had apparently been quite the contrary. Once upon a time, labouring human beings deployed their own skill and dexterity. The tool had been the means of transmission of the activity of the worker to its object. Under modern conditions, however, the machine has become the subject: 'Rather it is the machine which possesses skill and strength in place of the worker, is itself the virtuoso, with a soul of its own in the mechanical laws acting through it; and it consumes coal, oil etc. (*matières instrumentales*), just as the worker consumes food, to keep up its perpetual motion' (p. 693). It is as though the machine has had life

10 Marx, *The Communist Manifesto*, pp. 61–2.
11 Marx, *Grundrisse*, p. 692.

breathed into it – a spirit which sucks the vitality from the workers who are now mere appendages:

> The production process has ceased to be a labour process in the sense of a process dominated by labour as its governing unity. Labour appears, rather, merely as a conscious organ, scattered among the individual living workers at numerous points of the mechanical system; subsumed under the total process of the machinery itself, as itself only a link of the system, whose unity exists not in the living workers, but rather in the living (active) machinery, which confronts his individual, insignificant doings as a mighty organism. (*Ibid.*)

Marx offers a representation of domination and control – transferred energy, power, exploitative capacity – but no theorisation of the uncontrollable *per se*. Productive forces, power, destructive energy are always unleashed under the aegis of an historical project, a socially transforming energy system, above all capitalism.

To a large degree such points can also be made about Engels, although the balance is rather different. In *Anti-Dühring* (1878), he insists that changing economic relations produce shifts in the nature of warfare. War does not give birth to itself; its ultimate author is not force, but wider transformations in economy and society. Let us sketch out a little further his historical account of war. When, at the beginning of the fourteenth century, gunpowder came from the Arabs to Western Europe, this should be understood, we are told, as an effect of economic advance, progress in industry.[12] The introduction of firearms revolutionised the waging of war and this in turn affected the political relations of domination and subjection. The provision of powder and firearms required industry and money, which favoured the burghers of the towns. Indeed it was the towns and the rising monarchy (drawing its support from the urban centres) which effectively wielded the new arms against the feudal nobility. 'The stone walls of the nobleman's castles, hitherto unapproachable, fell before the cannon of the burghers, and the bullets of the burghers' arquebuses pierced the armour of the knights.'[13] Further advance was slow; it took over three hundred years before a weapon was constructed which was suitable for the equipment of the whole body of infantry, namely the flintlock musket with a bayonet, introduced in the early eighteenth century.

In Engels's view, the eighteenth century saw two further crucial developments: the American War of Independence, in which lines of infantry confronted highly motivated bands of insurgents and

12 See Engels, *Anti-Dühring*, p. 205.
13 See *ibid.*, pp. 205–6.

sharpshooters, and the French Revolutionary wars which consolidated what the American Revolution had begun and which
constituted the necessary precondition for socialism – the over-
throw of the old regime.[14]

But Engels argues that with the Franco-Prussian War less than
one hundred years later, technical advance came to a point of
climax and conclusion.[15] This was viewed as a culmination of past
developments and a turning point of entirely new significance.[16]
The weapons, he declares, had reached such a stage of perfection
that further revolutionising progress was no longer possible in
this sphere: all further improvements in field warfare would be
trivial. The era of evolution was therefore, in essentials, closed in
this direction. The next stage of change would be in the further
militarisation of society itself. Militarism, he grimly warned, was
swallowing Europe.[17]

Engels predicted that Prussia's military machine would produce
a war on a scale previously unimagined. To quote his famous
'uncanny' prediction in the 1880s of the coming world war:

And finally no war is any longer possible for Prussia-Germany
except a world war and a world war indeed of an extension and
violence hitherto undreamt of. Eight to ten millions of soldiers
will mutually massacre one another and in doing so devour the

14 Similarly in Lenin's *Socialism and War* (1915), the period from the French
    Revolution to the Paris Commune is perceived to be marked by a bourgeois
    progressive form of war against absolutism and feudalism. Whatever plunder
    or conquest took place, the historical significance of that period of warfare lay
    in destroying feudalism. Moreover whilst in the Franco-Prussian War
    Germany plundered France, this did not alter the significance of the war in
    liberating tens of millions of German people from feudal disunity and from
    the oppression of two despots, the Russian Tsar and Napoleon III. Only wars
    which were historically purposive and progressive could thus be justified – the
    position of say a slave-holder owning 100 slaves warring against another who
    owns 200 slaves for a more 'just' redistribution of slaves was quite different
    and could not be seen either as a defensive or as a progressive war; see Lenin,
    Socialism and War, p. 166.
15 See Engels, *Notes on the War*.
16 See Semmel, *Marxism and the Science of War*, p. 53.
17 See *ibid*., p. 54. Lenin would also argue in *Socialism and War* that what was
    formerly progressive had now become reactionary. Capitalism had developed
    the forces of production to such a degree that mankind is faced with the
    alternative of adopting socialism 'or of experiencing years and even decades of
    armed struggle between the "Great" Powers for the artificial preservation of
    capitalism by means of colonies, monopolies, privileges and national
    oppression of every kind. ... It is not the business of socialists to help the
    younger and stronger robber [Germany] to plunder the older and overgorged
    robbers' (*Socialism and War*, p. 167).

whole of Europe until they have stripped it barer than any swarm of locusts has ever done. The devastations of the Thirty Years' War will be compressed into three or four years, and spread over the whole Continent; famine, pestilence, general demoralisation both of the armies and of the mass of the people produced by acute distress; hopeless confusion of our artificial machinery in trade, industry and credit, ending in general bankruptcy; collapse of the old states ... absolute impossibility of seeing how it will all end and who will come out of the struggle as victor; only one result is absolutely certain: general exhaustion and the establishment of the conditions for the ultimate victory of the working class. This is the prospect when the system of mutual outbidding in armaments, driven to extremities, at last bears its inevitable fruits.[18]

That descent into war is portrayed as a *free-wheeling process*, irreducible to precise pre-given formulae or to political constraint. He acknowledges the absolute impossibility of seeing how it will all end. Modern war can only produce unknown results. As he puts it in the preface to a work of military history published in 1888: 'Peace continues only because the technique of armaments is constantly developing, consequently no one is ever prepared; all parties tremble at the thought of world war – which is in fact the only possibility – with its absolutely incalculable results.'[19]

Yet war is also grasped as dialectically purposive. Both visions exist in Engels's work: on the one hand war as a devouring force unpredictable in its outcome and subsequent effects; on the other hand war as a catalyst towards progress, a stage in an unfolding destiny which 'suits' his own calculations: 'And when nothing more remains to you [the old European statesmen] but to open the last great war dance – that will suit us all right.'[20] The expense of the army and navy, Engels argues, heralds impending financial catastrophe. War is a self-transforming and ultimately self-destroying process. The negation of peace by war is itself inevitably to be negated, engendering for Engels as later for Lenin the 'conditions for the ultimate victory of the working class'. The potential mobilisation and arming of entire peoples placed hitherto unprecedented power in the hands of the working classes, who might now sabotage the war machine in which they participated:

18   Marx and Engels, *Correspondence*, pp. 456–7.
19   Quoted in Gallie, *Philosophers of War and Peace*, p. 92.
20   Marx and Engels, *Correspondence*, p. 457. Marx and Engels thus read the Franco-Prussian War as purposive, furthering eventual liberation (via German victory); see Engels to Marx (15 August 1870), *ibid.*, p. 295.

At this point the armies of princes become transformed into armies of the people; the machine refuses to work, and militarism collapses by the dialectics of its own evolution. What the bourgeois democracy of 1848 could not accomplish, just because it was *bourgeois* and not proletarian, namely to give the labouring masses a will whose content was in accord with their class position – socialism will infallibly secure. And this will mean the bursting asunder *from within* of militarism, and with it of all standing armies.[21]

Such a purposive understanding is certainly evident in much Marxist thought on war. Lenin argues with regard to Clausewitz's dictum on war as the continuation of politics that 'Marxists have always rightly regarded this thesis as the theoretical basis of views on the significance of any war.'[22] The meaning and ramifications of war are to be located in the 'materialist' processes which give it birth and which constrain it. Thus for Lenin the plundering of colonies, the oppression of other nations and the suppression of the working-class movement is the sole political basis of the First World War.[23] Imperialism constitutes a method for averting civil war; expansion provides both resources and an ideological cohesion which masks internal contradictions. The network of finance expands globally although always unevenly: 'Thus finance capital, literally, one might say, spreads its net over all countries of the world.'[24] In this frame of reference, the First World War represents a struggle for 'booty' between European imperialist financiers. The war is the effect of a nineteenth-century economic trajectory, the movement towards the monopoly stage of capital. Lenin thus differentiates between socialism, bourgeois pacifism and anarchism. War cannot be abolished unless classes are abolished and the moment of socialism arrives.[25] It would then seem not only that wars are historically determined in their forms, but war as such has

---

21　Engels, *Anti-Dühring*, p. 210. *Cf.* Lenin in 1914: 'The proletarian banner of civil war will rally together, not only hundreds of thousands of class-conscious workers but millions of semi-proletarians and petty bourgeois, now deceived by chauvinism, but whom the horrors of war will not only intimidate and depress, but also enlighten, teach, arouse, organise, steel and prepare for war against the bourgeoisie of their "own" country and "foreign" countries. And this will take place, if not today, then tomorrow, if not during the war, then after it, if not in this war then in the next one' ('Position and tasks of the Socialist International', p. 40).

22　*Socialism and War*, p. 304.

23　See *ibid.*

24　Lenin, *Imperialism*, p. 63.

25　See *Socialism and War*, p. 304.

a beginning and an end in time. Proudhon, Marx, Engels and Lenin all refuse the supposed 'platitudes' of a simple 'bourgeois' pacifism. Yet each of them envisages the possibility of an eventual permanence of peace. Their position returns to a possibility of a final epoch of resolution, beyond war – what De Quincey, to whom I turn in the next chapter, had earlier dismissed as the 'most romantic of all romances'.

Even Rosa Luxemburg, for all her intense anguish at the outbreak of war in 1914 and its ambiguous consequences for 'this huge madhouse in which we are living', could sometimes comfort herself with the belief that the conflict would constitute, *by necessity*, the royal road to a progressive future, a triumph of the working class.[26] As she put it in an extraordinary letter to her friend Sonja Liebknecht in 1917:

> You know, Sonitschka, that the longer it takes, and the more the basenesses and atrocities occurring every day transgress all limits and bounds, the more calm and resolute do I become. Just as I cannot apply moral standards to the elements, to a hurricane, a flood, or a solar eclipse, and instead only consider them as something given, an object of research and knowledge, these are obviously, objectively, the only possible paths of history, and we must follow them without getting diverted from the basic direction. I have a feeling that this whole moral mire through which we are now wading, this huge madhouse in which we are living, can overnight, with the wave of a magic wand, be transformed into its opposite, transformed into something extraordinarily great and heroic and – if the war should last for another few years – it *must* be transformed. . . . We must take everything that happens in society the same way as in private life: calmly, generously and with a mild smile. I firmly believe that, in the end, after the war, or at the close of the war, everything will turn out all right. But apparently we must first wade through a period of the worst human suffering.[27]

26 On her eloquent perception of the Great War as an utter disaster, see the famous 'Junius Pamphlet'. She refers to the initial madness of 1914 enthusiasm, its 'delirous frenzy', and to the gathering atmosphere of tragedy – war's bleak chorus, 'the hoarse croak of the vultures and hyenas of the battlefield'. The war is viewed as both a catastrophe which has converted Europe into a cemetery, and as an inspiring 'turning point': 'Historically, the war is ordained to give to the cause of the proletariat a mighty impetus' 'The crisis in German Social Democracy (*The Junius Pamphlet: Part One*)', pp. 323, 329).

27 *The Letters of Rosa Luxemburg*, p. 235.

6 De Quincey's 'Most Romantic of All Romances'

To read Thomas De Quincey and John Ruskin on war is to encounter very different forms of social and political critique from those explored in the previous chapter, positions in which the vision of an encroaching spiritually impoverished capitalism is cast against the age of a 'true' romantic military contest. In each case modernisation, liberalism, *laissez-faire* and industrialisation are symptoms to be lamented. Both writers contrast modern forms of destruction with the sublime and chivalrous war of the past. The very soul of fighting, it seems, has been defiled and pitilessly mechanised.

On the other hand, we can also find links between the war argument of De Quincey and Proudhon. Whilst Proudhon's *War and Peace* has a certain self-consciously maverick, ironic and iconoclastic tone, it should not of course be seen as an isolated example of the intellectual advocacy of war in mid-nineteenth-century Europe. As new groupings, conferences and arguments for peace developed,[1] so also the critique of pacifism took on new forms with regard both to intra-European conflict and to imperial war.

The Crimean War occurred in the heyday of British ebullient liberal *laissez-faire* argument. It was cast by its opponents, such as Richard Cobden, as a grotesquely immoral and financially profligate adventure.[2] For many of its supporters, however, the war was a splendid antidote to contemporary commercial greed and sordid calculation, indeed the true regenerative route out of the fevered materialism of the age.[3]

The contest was perceived as a battle against insidious amoralism and corruption at home as well as against Tsarist autocracy and

1   The first London Peace Convention met in June 1843, the second in Brussels in 1848; the third in Versailles in 1849, presided over by Victor Hugo. In the same year, Cobden attempted to introduce a motion in Parliament proposing treaties which would ensure that international disputes were referred to arbitration. See Hopkins, 'De Quincey on war'.
2   See Chapter 2.
3   As Tennyson puts it in *Maud* (1855):
    So I wake to the higher aims
    Of a land that has lost for a little her love of gold,
    And love of a peace that was full of wrong and shames,
    Horrible, hateful, monstrous, not to be told;
    And hail once more to the banner of battle unroll'd!

Russian international brigandage. To support the 'infidels' clearly involved complex convolutions of 'Christian' rhetoric, but the Turks were now reassuringly declared to be liberalising and moralising their own once execrated regime. If the Crimean War witnessed an ideological divide between peace and war advocates, it also constituted a technological watershed: a crossing point between sail and steam, providing a demonstration of a new naval mobility and power in the absolute destruction of the Turkish squadron at Sinope.

Thomas De Quincey's essay, 'On War' (published in 1854 at the time of the Crimean War but probably written significantly earlier), is far less well known than say Tennyson's poetics of battle. It rejects the 'Christian' idea of abolishing military conflict as 'the most romantic of all romances'.[4] As John Barrell has recently shown, 'the East' was to be the particular and infinitely complex locus for De Quincey's terrors and horror; the domain above all others on which it was discursively and militarily essential to make war. In his essays on Sino-British relations, for instance, De Quincey makes it clear that he approves of pitiless military retribution in cases of Chinese transgression – or indeed in anticipation of future crimes and insubordination. Where Cobden protests the iniquity of Britain's savage repression in Canton and elsewhere in the East, De Quincey sees force as essential against 'our hateful enemy', the Chinese.[5]

Unlike so many of the other writers I discuss, De Quincey's primary concern is indeed making war with the non-European world. Intra-European war anxiety rather than the ideology of colonisation and imperial war is my main area of focus in this book, but even where the explicit defence fears or war rationales I investigate below concern the French or German threat, other more distant battles or 'universal laws of conquest', resentful 'half-caste' spies or unreliable 'alien races in our midst' often intrude. Many of my commentators detect a complex, atavistic and supposedly non-European racial descent beneath the skin of their continental foe.

In 'On War', De Quincey witheringly disparages 'the flying leap' by which the idealist seeks to clear 'this unfathomable gulf of

---

4 De Quincey, 'On War', p. 195. A version of the essay may have been written as early as the 1830s or 1840s. For De Quincey's interest in Kant, notably *Perpetual Peace* see Hopkins, 'De Quincey on war'.

5 On Cobden, see Chapter 2. On De Quincey and the East, see Barrell, *The Infection of Thomas De Quincey*, pp. 150, 155. I cannot rehearse Barrell's intricate argument here. I focus rather on the abstracted figure of conflict in De Quincey's essay on war – a text which, curiously, Barrell does not mention.

war and to land his race forever on the opposite shore of a self-sustaining peace' (p. 216). There is, he declares, a twofold necessity for war. First, there is 'A physical necessity arising out of man's nature when combined with man's situation – a necessity under which war may be regarded, if you please, as a nuisance, but as a nuisance inalienable from circumstances essential to human frailty.' But secondly, there is 'A moral necessity connected with benefits of compensations, such as continually lurk in evils acknowledged to be such – a necessity under which it becomes lawful to say that war *ought* to exist as a balance to opposite tendencies of a still more evil character' (pp. 200–1). His train of thought becomes still more startling in the next formulation:

> War is the mother of wrong and spoliation; war is a scourge of God. Granted; but, like other scourages in the divine economy, war purifies and redeems itself in its character of a counterforce to greater evils that could not otherwise be intercepted or redressed. (*Ibid.*)

Once again war is gendered there; it is the mother, more specifically the mother of 'wrong and spoliation'. War engenders harm, but it is also a self-purifying and redeeming force in 'the divine economy'. The fact that there is no imaginable machinery which could abolish war is 'on the whole a blessing from century to century, if it is an inconvenience from year to year' (pp. 202–3). It neither can nor ought to be abolished. At best, it can be hoped that a widening gap may emerge between the mechanisms of diplomacy and the final resort to war.[6] The very idea of a war on war is fallacious; or rather, war is naturally an antidote to war, since, like a pendulum, the world is destined to swing back and forth between states of war and peace.[7] This perception of war is part of De Quincey's wider understanding of the conflict-ridden nature of being and knowledge. For De Quincey thought itself only occurs through antagonism. Moreover only through the loss of a certain state do we come to grasp, retrospectively, what that state meant. The world is constituted in the play of agencies and counter-agencies, actions and reactions.[8]

6 'Although war may be irreversible as the last resource, this last resource may constantly be retiring farther into the rear' (p. 219).

7 'War has no tendency to propagate war, but tends to the very opposite result' (p. 220); 'a new war as certainly becomes due during the evolutions of a tedious peace as a new peace may be relied on during the throes of a bloody war' (p. 212). As things stand at present, war and peace are bound together like the vicissitudes of day and night' (p. 211).

8 See Miller, *The Disappearance of God*, ch. 2.

Wars are caused, De Quincey suggests, by the system of 'national competitions', not by superficial causes and catalysts ('proximate excitements'); state conflict cannot be superseded since it is inevitably the final court for the settlement of disputes: 'All war is an instinctive *nisus* for redressing the errors of equilibrium in the relative positions of nations amongst nations' (pp. 210–11). The rivalry between nations is deemed to be as natural as that between wild beasts. All nations have to watch vigilantly that they are not losing ground in the international competition. The ambassador inevitably speaks with one hand on the sword whilst diplomacy 'is the graceful drapery which shrouds their [nations'] natural, fierce, and tiger-like relations to each other' (p. 211). War cannot be banished, although it could, potentially, be displaced backward from its present form into the conflicts typical of an earlier ruder stage of civilisation: 'Banish war as now administered, and it will revolve upon us in a worse shape, that is, in a shape of predatory and ruffian war, more and more licentious' (pp. 212–13).

War between states is a preferable substitute for that more anarchic war of all against all which marked the past: thus paradoxically the 'civilised' Christian gesture of seeking the abolition of modern war ('wars of high national police, administered with the dignified responsibility that belongs to supreme rank', p. 214) would in fact produce a terrible regression ('an interminable warfare of a mixed character . . . infesting the frontiers of all states like a fever'). Christian interference would promote the return to some 'lawless *guerrilla* state' of war, reducing society to primitive factionalism and fostering 'precisely the retrograde or inverse course of civilization' (*ibid.*). What we encounter here is an argument which, as we shall see later, was to be entertained by Freud: that war reveals the state's attempt to mobilise and monopolise the violence dispersed within it.[9] In the 'natural order of civilization', De Quincey suggests, war passes from the hands of 'knights, barons and insulated cities into those of the universal community' (*ibid.*). In short, war becomes ever more rational and civilised because 'continually more intellectual'; it opens 'into wide scientific arts, into strategies, into tactics'. War, he insists, is ever more powerfully imbued with 'the exquisite resources of science'. It is continually refining itself 'from a horrid trade of butchery into a magnificent and enlightened science' (p. 230).

Elsewhere De Quincey writes in a less sanguine tone of science and technological advance. In 'The English Mail Coach; or the

9 See Chapter 15.

Glory of Motion' (1849), he laments the passing of the age of the mail coach which represented a world of spatial connectedness; a society in which speed was related to animal energy and was humanly comprehensible: 'This speed was incarnated in the *visible* contagion amongst the brutes of some impulse that, radiating into *their* natures, had yet its centre and beginning in man.'[10] In the age of Trafalgar and Waterloo, news was still transmitted across the country by mail coaches, 'like fire racing along a train of gunpowder . . . kindling at every instant new successions of burning joy' (p. 156). The coach moved forward at human speed, speaking 'the language of our victorious symbols', 'our martial laurels' and setting in train 'a grand national sympathy' (pp. 157–8).

De Quincey offers an account of past glory and present mechanical degradation which insists that the highest truth is embodied in the fervent story of 'the sublime regiment' and its heroic fall. Here follows the noble narrative to be told to the suffering mother awaiting news of her soldier son:

> I lifted not the overshadowing laurels from the bloody trench in which horse and rider lay mangled together. But I told her how these dear children of England, privates and officers, had leaped their horse over all obstacles as gaily as hunters to the morning's chase. I told her how they rode their horses into the mists of death, (saying to myself, but not saying to *her*) and laid down their young lives for thee, O mother England! (p. 161)

Yet that world of higher truths is threatened. For now, 'the new system of travelling, iron tubes and boilers' has disconnected man's heart from 'the ministers of his locomotion' (p. 143):

> The galvanic cycle is broken up for ever: man's imperial nature no longer sends itself forward through the electric sensibility of the horse; the inter-agencies are gone in the mode of communication between the horse and his master, out of which grew so many aspects of sublimity under accidents of mists that hid, or sudden blazes that revealed, of mobs that agitated, or midnight solitudes that awed. (*Ibid.*)

Modern technology profoundly deranges social relationships and the connection between the human worker and the tool – De Quincey would agree with Marx and Engels there, although his language and his conclusions are different. The harnessing of 'vast physical agencies', 'steam in all its applications', daguerreotypes,

---

10 De Quincey, 'The English Mail Coach; or the Glory of Motion', p. 143.

education, the accelerating power of the press and new military technology are at once sublime and terrible, he insists in *Suspiria De Profundis* (1845). Modernity disconnects the means of communication from 'man's imperial nature'. In face of these 'powers from hell (as it might seem, but these also celestial) coming round upon artillery and the forces of destruction – the eye of the calmest observer is troubled':

> the brain is haunted as if by some jealousy of ghostly beings moving amongst us; and it becomes too evident that, unless this colossal pace of advance can be retarded, (a thing not to be expected,) or, which is happily more probable, can be met by counter-forces of corresponding magnitude, forces in the direction of religion or profound philosophy, that shall radiate centrifugally against this storm of life so perilously centripetal towards the vortex of the merely human, left to itself, the natural tendency of so chaotic a tumult must be to evil; for some minds to lunacy, for others to a reagency of fleshly torpor.[11]

In the modern age, 'the eye of the calmest observer is troubled'. Yet De Quincey's recoil from modern machinery and his advocacy of 'the machinery of dreaming' is not allowed to affect his stark argument on war. To reject the inevitability of war, it seems, is merely abstract piety. Conflict is a fact of the human and animal condition in De Quincey's account, although he lacks any developed theory of the function of that conflict – the language of war which evolutionary naturalism would so strikingly develop in the later part of the century.[12] Beyond De Quincey's or Proudhon's romance of war lay the crisis of the Franco-Prussian conflict and the Paris Commune in 1871, the year of Darwin's *Descent of Man*.

11   De Quincey, *Suspiria de Profundis*, p. 148.
12   See Chapter 8.

7 Ruskin and the Degradation of True War

The very meteorology of the late nineteenth century, it seemed to John Ruskin, changed after the Franco-Prussian War. *The Storm Cloud of the Nineteenth Century* (1884) describes a phenomenon which in recent years had been intermittently hovering over England, brought by a wind of darkness, pollution and malignancy – 'one loathsome mass of sultry and foul fog', a 'sulphurous chimney-pot vomit of blackguardly cloud'.[1] Ruskin locates a new climatic, industrial and national event, a terrible 'sign of the times', of industry and war, of chimneys, furnaces, guns, materialism and blindness: 'Blanched Sun, – blighted grass, – blinded man' (p. 61). Like Dickens's fog on the first page of *Bleak House* (1852–3), Ruskin's cloud implies outrageous human-induced confusion and pollution. The sky looks deadly; it is shrouded by weird grey cloud; not rain-cloud, but a 'dry black veil, which no ray of sunshine can pierce'.

For Ruskin in the 1880s (unlike Dickens in *Bleak House*), the cloud image also specifically connects with war. The social critique and war critique converge. He is able to date the cloud to the year 1871, and more specifically to the early spring of that year. As he puts it in a passage from *Fors Clavigera* (1871–84), reproduced in the *Storm Cloud* lecture: 'It looks to me as if it were made of dead men's souls – such of them as are not gone yet where they have to go, and may be flitting hither and thither, doubting themselves, of the fittest place for them' (p. 48). Glossing his own quotation, Ruskin indicates that the sentence refers to 'the battles of the Franco-German campaign, which was especially horrible to me, in its digging, as the Germans should have known, a moat flooded with waters of death between the two nations for a century to come' (*ibid.*).

Ruskin's judgement of political economy turns on the distinction between life and death values, human and mere money wealth. The

---

1 Ruskin, *Storm Cloud*, pp. 55, 57. Note that Ruskin's Brantwood period began in 1871 when just after his delirious illness at Matlock, he purchased a house overlooking Coniston Water. There was indeed an exceptional period of high rainfall, extreme cold and little sunshine during the 1870s: see Rosenberg, *Darkening Glass*, pp. 201, 214.

issue of war profiteering, for instance, highlights the disparity between different criteria of worth. As he puts it in *Fors Clavigera*:

> You, or your fellows, German and French, are at present busy in vitiating [the air] to the best of your power in every direction; chiefly at this moment with corpses, and animal and vegetable ruin in war: changing men, horses, and garden stuff into noxious gas. But everywhere, and all day long, you are vitiating it with foul chemical exhalations; and the horrible nests, which you call towns, are little more than laboratories for the distillation into heaven of venomous smokes and smells, mixed with effluvia from decaying animal matter, and infectious miasmata from purulent disease.[2]

European wars today, Ruskin insists, are to do with thieving and coveting each other's land, or, to state it more generically: 'the real sources of all deadly war . . . are capitalists' (p. 127). Where Cobden had seen capitalism as fundamentally peaceful, Ruskin suggests its fatal relationship to war.

The *Storm Cloud* refers to vaster processes of which the war is only a symptom: a century of industry and empire which produces pollution and death: 'the Empire of England, on which formerly the sun never set, has become one on which it never rises'. In other words, the idea of the lamps going out in war was not conjured out of the blue on the eve of the First World War in the British Foreign Secretary Sir Edward Grey's famous remark, but has a place in a longer history of concerns with diminishing light and fading power, above all the prospect of the death of the sun itself. Thermo-dynamics involves the exploration of the relation between heat and mechanical energy; the second law declares the impossibility of producing work through the transfer of heat from a cold body to a hot body in any self-sustaining process. Stated in another way, entropy always increases in any closed system not in equilibrium. Such a conception of increasing disorder, irreversibility and depletion of power in systems not in equilibrium was to be interpreted and cross-referenced to social change in many ways. Certainly for some it suggested a bleak and imminent future of cold, darkness and depletion. On the basis of the second law, Lord Kelvin had quickly reached toward the cosmic implications of eventual sun death in the 1850s (as does Dickens on the first page of *Bleak House*), providing a severe jolt to ebullient Victorian predictions of infinite human, global and cosmic progress.[3] What emerged was a way of reading

2   *Fors Clavigera*, p. 91.
3   See Brush, *The Temperature of History*.

industry, machinery and technology which converges with the massively disturbing experience of 1870–1. In addition to the critique of the triumphalist march of the machine and its attendant brutalities and 'hard times', or the lament for the nexus of technology, industry and 'progress' in Carlyle and more generally the early Victorian 'condition of England' novel, a further discursive possibility emerges: machinery and machine mentality as the source of war, suicidal heat loss, fatal friction, degeneration, the inevitable running down of the system. Ruskin is not simply rehearsing the language of early or mid-Victorian social protest against the machine. He could be said to provide the benchmark of that intensified discursive relationship between technology, trade and destruction at issue throughout the present investigation. As he put it in 1871:

> there is no physical crime to this day, so far beyond pardon, – so without parallel in its untempted guilt, as the making of war-machinery, and invention of mischievous substance. Two nations may go mad, and fight like harlots – God have mercy on them; – you, who hand them carving knives off the table, for leave to pick up a dropped six pence, what mercy is there for *you*?[4]

For Ruskin, the destructive, atomising propensities of modern machinery are so awful that ultimately it is necessary to repudiate industrialisation as such (unlike Carlyle and Dickens who call rather for its moralisation). The body and the landscape are characteristically to be viewed as either utterly pure or impure. The same is also true of war, he suggests. Industry and political economy are seen as a physical and moral desecration of nature's loveliness. Take the first of the lectures which compose *The Crown of Wild Olive* (1866). Ruskin writes of a pure source of water corrupted by an unthinking humanity; the lovely spot lies in the south of England bordering on the sources of the Wandel: 'No clearer or diviner waters ever sang with constant lips of the hand which "giveth rain from heaven".'[5] These waters are insolently defiled by 'the human herds that drink of them' (p. 385). Ruskin's extraordinary prose wells up, cascading through the water scene it recounts:

> Just where the welling of the stainless water, trembling and pure, like a body of light, enters the pool of Carshalton, cutting itself a radiant channel down to the gravel, through warp of feathery weeds, all waving, which it traverses with its deep threads of

4 *Fors Clavigera*, p. 130.
5 *The Crown of Wild Olive*, p. 385.

clearness, like the chalcedony in moss-agate, starred here and there with the white grenouillette; just in the very rush and murmur of the first spreading currents, the human wretches of the place cast their street and house foulness; heaps of dust and slime, and broken shreds of old metal and rags of putrid clothes; which, having neither energy to cart away, nor decency to dig into the ground, they thus shed into the stream, to diffuse what venom of it will float and melt, far away, in all places where God meant those waters to bring joy and health.... the clean water nevertheless chastises to purity; but it cannot conquer the dead earth beyond: and there, circled and coiled under festering scum, the stagnant edge of the pool effaces itself into a slope of black slime, the accumulation of indolent years. (p. 386)

In one sense, the environment is radically divided between good and evil, purity and pollution; but the problem is precisely one of seepage; pools, flowers, summer air, river banks, 'glittering waves medicinal' are counterposed to and yet infiltrated by filth, slime and mud – the processes of neglect and stagnation.[6] Just as the waters must be cleansed of the filth that invades them, so the interminable and perhaps hopeless social task is to purify the nation, and redeem the original truth of war: 'Make your national conscience clean, and your national eyes will soon be open' (p. 478). For Ruskin the path to that purity lies not through class conflict, but a kind of socio-moral surgery – a radical probing and differentiation of knights and parasites, workers and idlers, honest souls and fiends.

In 'War', the third of the lectures which comprise The Crown of Wild Olive,[7] Ruskin insists that national military conflict is not anathema to art, but its very basis. Lest he be accused of abstracting carnage into mere conjecture and rumination, he acknowledges his own painful personal experience of war, deprived by Austrian shells of three of his favourite paintings by Tintoretto. Nevertheless war is the foundation of art:

6   See ibid., p. 386.
7   The Crown of Wild Olive consists of four lectures: 'Work' [1865], given at the Working Men's Institute, Camberwell; 'Traffic' [1864] delivered at the Town Hall, Bradford; 'War' [1865] presented at the Royal Military Academy, Woolwich; and 'The Future of England' [1869] addressed to the Royal Artillery Institute, Woolwich. In 1866 Ruskin had collected the first three lectures (much revised) into a volume; in 1873 he added to the three lectures a fourth, and other material including an appendix, 'The Political Economy of Prussia'. It seems that there is no record at Woolwich of Ruskin's lecture; nor apparently were there any press notices; see Rosenberg, Darkening Glass, p. 158.

all the pure and noble arts of peace are founded on war; no
great art ever yet rose on earth, but among a nation of soldiers.
There is no art among a shepherd people, if it remains at peace.
There is no art among an agricultural people, if it remains at
peace. Commerce is barely consistent with fine art; . . . There is
no great art possible to a nation but that which is based on battle.
(p. 459)

Much ancient art, he points out, is directly concerned with the
representation of war. Peace, moreover, apparently produces
cultural decadence. Tracing the genius of the Egyptians and the
Greeks, the artistic poverty of the Romans and the Gothic revival
through to the great heights of the Renaissance in Lombardy and
Tuscany, he finds a symbiotic relationship between war and art, for
'as peace is established or extended in Europe, the arts decline' (p.
463). For Ruskin, as earlier for Hegel, De Quincey and Proudhon,
war is perceived as indispensable to the survival of 'culture' and as
the very foundation of the nation – hence he speaks of the 'creative,
or foundational, war' (p. 465).

Military conflict supposedly produces both virtue and art, or
rather the kind of true creativity which is inseparable from virtue:

> We talk of peace and learning, and of peace and plenty, and of
> peace and civilisation; but I found that these were not the words
> which the Muse of history coupled together: that, on her lips, the
> words were – peace, and sensuality – peace, and selfishness –
> peace, and death. (p. 414)

But Ruskin also recognises war as the scene of a contradiction. It is
impossible, he suggests, to avoid reproducing that contradiction
in seeking to represent war. It conjures up both the image of a
pure, lost tournament and the field of modern military-industrial
degradation. 'It is impossible for me to write consistently of war,
for the group of facts I have gathered about it leads me to two
precisely opposite conclusions' (p. 515). Since it is the site of incalculable human suffering, war perhaps ought to end in Christian
nations. Nevertheless, 'the most beautiful characters yet developed
among men have been formed in war' (*ibid.*).

The point is not that any war will achieve Ruskin's ends; some
conflicts are trivial, ignoble or simply irrelevant. The real concern is
not with the wars of 'barbarism' or of mountain clans, nor wars of
ambition and power (in which category he includes Napoleon's
wars and the American Civil War) but with wars that represent a
decisive playing out of a noble human drama, effects of 'the natural
restlessness and love of contest among men [who] are disciplined,

by consent, into modes of beautiful – though it may be fatal – play'
(p. 465). War is noble precisely when it is conscious – a battle of
the masters and not the slaves.[8]

Noble war has to be the direct expression of male pugnacity: 'All
healthy men like fighting, and like the sense of danger; all brave
women like to hear of their fighting, and of their facing danger.
This is a fixed instinct in the fine race of them' (p. 469). Women
provide not only war's audience, but also its catalyst – as Ruskin
notoriously declares: 'you tender and delicate women, for whom,
and by whose command, all true battle has been, and must ever be'
(p. 466); or on another occasion, in *Sesame and Lilies* (1865),
'There is not a war in the world, no, nor any injustice, but you
women are answerable for it; not in that you have provoked, but
that you have not hindered'.[9]

It perhaps goes without saying that no actual war (nineteenth-
century or otherwise) bore much relation to Ruskin's utopian
vision. But such conceptions nevertheless had their historical importance in providing a perspective for the understanding of conflict
in the period – ideal codes of honour, chivalry, nobility – against
which the shocking discrepancy of the real could be measured.
Moreover, even the Franco-Prussian War could be conceived, or
rather caricatured, as a 'duel' between two individuals. Figure 3
from *Punch* depends for its effect on such a notion of war as
individuated conflict between two grand figures.

Death in battle would, ideally, provide an index to valour and
strength since, if war is properly 'played', the noblest survives. The
reference here is not specifically to Darwinism, but it takes for
granted a conception of natural selection:

> First, the great justification of this game [war] is that it truly,
> when well played, determines *who is the best man*; who is the
> highest bred, the most self-denying, the most fearless, the coolest
> of nerve, the swiftest of eye and head. You cannot test these

8   Ruskin cites Carlyle from *Sartor Resartus* (Bk II, ch. viii) on the absurdity of
    the situation during the Napoleonic wars where useful craftsmen from British
    villages confronted their French counterparts in the south of Spain with whom
    they had no quarrel. The conflict was between their masters alone. Yet at the
    order 'fire', they turned themselves into carcasses. See Carlyle, *Sartor Resartus*,
    pp. 119–20. *Cf.* 'capitalists, when they do not know what to do with their
    money, persuade the peasants in various countries, that the said peasants want
    guns to shoot each other with. The peasants accordingly borrow guns, out of
    the manufacture of which the capitalists get a percentage, and men of science
    much amusement and credit' (Ruskin, *Munera Pulveris*, p. 142).
9   Ruskin, *Sesame and Lilies*, p. 140.

A DUEL TO THE DEATH.

France. "PRAY STAND BACK, MADAM. YOU MEAN WELL; BUT THIS IS AN OLD FAMILY QUARREL, AND WE MUST *FIGHT IT OUT!*"

3   'A Duel to the Death' (from *Punch*, 23 July 1870).

qualities wholly, unless there is a clear possibility of the struggle's ending in death.[10]

War had once been sublime, a field in which 'Every man put on a crown, when the band of flute-players gave the signal for attack; all the shields of the line glittered with their high polish, and mingled their splendour with the dark red of the purple mantles' (p. 473). The rottenness and evil within the combatants was exposed and burnt away. But war has now been degraded, corrupted by the industrial revolution. This mechanisation of war is anathema to its higher meaning:

> You must not make it the question, which of the combatants has the longest gun, or which has got behind the biggest tree, or which has the wind in his face, or which has gunpowder made by the best chemists, or iron smelted with the best coal, or the angriest mob at his back. (p. 471)

That would be mere slaughter, iniquity and 'machine-contriving' instead of 'pure trial'.[11] War has been ruined, perhaps fatally; democracy ('our national crisis', p. 503) has vitiated the hierarchical order of the knights. Science itself is both agent and symptom of modern decline: 'the progress of science cannot, perhaps, be otherwise registered than by facilities of destruction' (p. 472). And yet war must still be hallowed, its truth and purity earnestly prayed for: 'Yet hear, for a moment, what war was, in Pagan and ignorant days; – what war might yet be, if we could extinguish our science in darkness, and join the heathen's practice to the Christian creed' (p. 472).

Ruskin's distaste for the scientific and mechanical war is part of his wider criticism of nineteenth-century socio-economic development. 'Engine-turned precision' is incompatible with the humanity of labour. He would do away with the railways in Wales and most of England, and he would destroy new towns, he declares in *Fors Clavigera* (p. 15).

In an appendix to the 1873 edition of *The Crown of Wild Olive* entitled 'Notes on the Political Economy of Prussia', Ruskin admits

10 *Crown of Wild Olive*, p. 470.
11 Truly noble war lies between the barbaric and the scientific phases. The problem is not war in itself, but its degradation through science. In short, the valid critique would not be of the old chivalric ethos of war, but only of modern war, the 'destructive machines' which ravage the country of the enemy and kill thousands: 'That, I say, is *modern* war, – scientific war, – chemical and mechanic war, – how much worse than the savage's poisoned arrow!' (p. 472).

that the relation between war and the Prussian tradition raises some vexing questions, although he endorses Carlyle's celebration of Frederick as the ideal king, who always thought of his people and recognised the necessity of both peace and war. Only soldiers have hitherto proved suitable for kingship, Ruskin argues. The king must be in a position personally to risk and to inflict death.

The notes are a paean to the greatness of the Hohenzollerns, to their painstaking construction of the state and to their creation of an ultimately inextinguishable national virtue. But here an enigma emerges, a crucial puzzle of how such racial virtue is born and nurtured in different settings. Ruskin puzzles over the conditions needed to retain all those putatively good Northern racial qualities through the tribulations of war and peace:

> The work now chiefly needed in moral philosophy, as well as history, is an analysis of the constant and prevalent, yet unthought of, influences, which, without any external help from kings, and in a silent and entirely necessary manner, form, in Sweden, in Bavaria, in the Tyrol, in the Scottish border, and on the French coast, races of noble peasants; pacific, poetic, heroic, Christian-hearted in the deepest sense, who may indeed perish by sword or famine in any cruel thirty years' war, or ignoble thirty years' peace, and yet leave such strength to their children that the country, apparently ravaged into hopeless ruin, revives, under any prudent king, as the cultivated fields do under the spring rain. How the rock to which no seed can cling, and which no rain can soften, is subdued into the good ground which can bring forth its hundred fold, we forget to watch, while we follow the footsteps of the sower, or mourn the catastrophes of storm. All this while, the Prussian earth, – the Prussian soul, – has been dealt upon by successive fate; and now, though laid, as it seems, utterly desolate, it can be revived by a few years of wisdom and of peace.[12]

To pose the relationship of war, revival and nationhood in the early 1870s, even for Carlylean Germanophiles was perhaps inevitably to hesitate, to face a deep confusion – the felt inadequacy of the appeal to 'races of noble peasants; pacific, poetic, heroic'. As Ruskin admits:

> And I have been hindered from completing my long intended notes on the economy of the Kings of Prussia by continually

12 *Crown of Wild Olive*, appendix, p. 529.

increasing doubt how far the machinery and discipline of war, under which the honesty and sagacity of Friedrich who so nobly repaired his ruined Prussia, might have done for the happiness of his Prussia unruined. (p. 516)

8 The Biology of War

Evolutionary theory was increasingly to be invoked in later nineteenth-century accounts of the necessity of war. Inter-state conflict, it seemed, was the brutal but necessary social equivalent to the 'natural struggle', the indispensable method for sorting the weak from the strong. Moreover, war was often cast as the guarantor of a certain natural, biological or even racial progress.

The 1850s and 1860s produced a variety of anthropological formulations about racial inequality and war. The French race theorist Arthur de Gobineau, whose *Essay on the Inequality of the Human Races* (1853–5) was little read at the time but rediscovered after 1870,[1] gave extreme expression to the view that historical degeneration was the effect of racial miscegenation, and war was the moment where the weakness of the diluted race was exposed. Vigorous invasion and the annihilation of the defeated were to be understood as the ineluctable and natural triumph of the racially strong on the one side, the necessary capitulation of the degenerate stock on the other:

> Societies perish because they are degenerate. . . . The word degenerate when applied to a people means . . . that the people had no longer the same intrinsic value as it had before, because it has no longer the same blood in its veins, continual adulterations having gradually affected the quality of that blood.[2]

According to Gobineau, historical degeneration stems from the increasing impurity of a race's blood. He rejects all theories which hold that moral corruption, religious fanaticism, economic decline or anachronistic political institutions are the primary agents of the fall of nations. The sole fundamental reason for social degeneration and military defeat is racial decay resulting from the vitiating mixture of superior and inferior races. Teutonic invaders produced the greatness of France whilst its gradual decline resulted from Celtic and Roman racial stock in the South. The highly centralised form of government, Roman in origin, gradually broke the influence of the feudal aristocracy, he argues, and with the Revolution of 1789 the

1   See Michael Biddis's introduction to Gobineau, *Selected Political Writings.*
2   Gobineau, *Political Writings*, pp. 58–9.

anarchistic South took its full revenge. In the light of history, Gobineau found himself:

> gradually penetrated by the conviction that the racial question overshadows all other problems of history, that it holds the key to them all, and that the inequality of the races from whose fusion a people is formed is enough to explain the whole course of its destiny. (p. 41)

The enigma and the tragedy of civilisation lies in the inevitable dilution of Aryan with other bloods. Collective demise, it seems, is as inevitable a fate as individual death:

> The fall of civilisations is the most striking, and, at the same time, the most obscure of all the phenomena of history...every assemblage of men, however ingenious the network of social relations that protect it, acquires on the very day of its birth, hidden among the elements of its life, the seed of an inevitable death. (pp. 42–3)

Gobineau's work combines racial anthropological jargon and a mystical sense of social destiny. It conceives of a racial dialectic of masters and slaves, and presents it by and large in naturalistic terms.

In the second half of the century both peace and war arguments were increasingly to be grounded in the authority of biology in general, and evolutionary arguments in particular. Patrice Larroque, author of a study of war in 1856, for instance, observed under the heading 'Debasement of the Human Species caused by War', that the removal of the vast majority of soldiers from the sphere of marriage and reproduction was biologically very worrying since it encouraged the proliferation of the weaker non-warrior population. In short, war furthered the degeneration of the species.[3]

The notion of a severe racial cost incurred in war was further elaborated and debated during the later part of the century. In 1895, to take a further example, Dr Charles-Jean Letourneau, a writer and authority on racial anthropology, criticised apologists from De Maistre to Proudhon who either through a misunderstanding of religion or through 'sophisms' argued that war was indispensable to progress. War, he argued, was bound up with evolutionary regression into savagery not progress.[4] Even military selection and recruitment, let alone the mass killing of war, had a

---

3   Larroque, *De la Guerre*, pp. 217–18.
4   Letourneau, *La Guerre dans les diverses races humaines*, p. 532.

regressive effect on human beauty, health and vigour. 'Military selection is necessarily retrograde since it exposes the flower of the species to death and disease'; 'physical deterioration inevitably ensues from this reverse selection' (p. 548). In other words, advocacy of war as the rightful site of natural selection between nations was countered with the argument that war does away with the fittest. The best-known exponent of this anti-militaristic view of evolutionary development was undoubtedly Herbert Spencer.

In his *Principles of Sociology* (1876–96), Spencer sought to generalise about what he called 'militant' and 'industrial' societies in order to produce a globally valid theory of social evolution. Like Richard Cobden, Spencer argued that war had been (or at least ought to have been) surpassed in the progress of history. He acknowledged that primitive societies were founded on the war ideal, but claimed that this was eventually surpassed: 'In rude societies all adult males are warriors; consequently, the army is the mobilized community, and the community is the army at rest'.[5] In the normal course of change the military leader grew into the political head, and the military function in society was separated from and subordinated to the supreme civil authority. It became a specialised tool of the society, no longer co-extensive with it. A system where land, for instance, constituted a 'trophy' for the victors was thus gradually and naturally displaced in the history of civilisation into a system of freely owned and traded property.

The shift from nomadic to settled life naturally produced a growing resistance to military service and action, Spencer went on. In their progressive evolution, societies ceased to be communities of 'hunters'. Spencer divided society into two forms – the 'militant' and 'industrial' types. Despite their commingling in practice, he believed that they displayed an opposite character in principle. The industrial form was thus, in its essence, opposed to war. In the militant society (exemplified by Russia and Prussia), the state appropriates industry, above all railways for military functions. Political control gradually slides back towards the army. In the industrial society (such as Britain) on the other hand, the historical trajectory is towards free trade, liberalism and a gradual decline in the role of the state. The industrial society is by definition anti-despotic, we are told, and hostile to all collectivist enterprises whether in the form of reactionary militarism, socialism or communism. It depends upon an 'industrial struggle for existence' in which competition and individualism are central. In Spencer's version of the classical liberal tenet of liberty: 'Omitting criminals

5 *Principles of Sociology*, vol. II, p. 563.

(who under the assumed conditions must be very few, if not a vanishing quantity), each citizen will wish to preserve uninvaded his sphere of action, while not invading others'.[6] The state withers, violent crime recedes and nationalism evaporates in an increasingly interdependent trading world.

But what troubled Spencer's picture of progress was the industrial society's 'transgression of the natural path' he had laid down for it. He witnessed a powerful backsliding from 'progress', even in Britain. In the terms of his naturalistic vocabulary, this could only be characterised as atavistic, a 'reversion' to an earlier 'type' of society.[7] Indeed the natural-evolutionary guarantee that progress *must* occur in human history, of which Spencer had so confidently reassured his reader around the mid-century,[8] became increasingly elusive in his later work. He anxiously charted the revival of military activity in Britain. The intensifying tendency of newspapers to revel in the details of bloody conflicts, the assimilation of volunteer forces to the regular army, proposals for their (offensive) deployment abroad, the growing tendency for officers to wear uniform even when off duty ('as they do in more militant countries'),[9] even the fact that the police now wore helmet-shaped hats: all of these features constituted an alarming regression. From the propensity of the executive to override other government agencies to the suggestion that all authorised teachers should be registered; from the idea of central inspection of local libraries to schemes for compulsory insurance, Spencer found many signs of collapse from an 'industrial' to a 'militant' society. Thus by 1882, when the second volume of *The Principles of Sociology* appeared, he had evidently come to doubt the natural inevitability of peaceful development upon which his social philosophy had hitherto been based, and upon which belief his present work was constituted. The future could now go either way:

> On the one hand, in the present state of armed preparation throughout Europe, an untoward accident may bring about wars which, lasting perhaps for a generation, will re-develop the coercive forms of political control. On the other hand, a long peace is likely to be accompanied by so vast an increase of manufacturing and commercial activity, with accompanying growth of the appropriate political structures within each nation, and

6 *Ibid.*, p. 702.
7 See *ibid.*, vol. II, p. 567.
8 For a discussion of Spencer's optimistic views about the inevitability of progress, see Peel, *Herbert Spencer*.
9 *Principles of Sociology*, vol. II, p. 681.

strengthening of those ties between nations which mutual dependence generates, that hostilities will be more and more resisted and the organization adapted for the carrying them on will decay. (vol. II, p. 736)

An elaborate range of evolutionist war and peace rationales developed in the later Victorian and Edwardian period. A full study of the nexus between evolutionary theory and the perception of violence, imperial conquest and war, or indeed of debates about the particular military strengths and weaknesses of colonial troops (for instance Mangin's perception that a black army, *La Force Noire*, 1910, should be used for offensive infantry attacks because of the natural savagery of such non-European troops)[10] is beyond my scope here. My purpose in dwelling upon these evolutionist arguments for and against war in the present chapter is to draw attention to common elements of the opposing views. It is the terms of this constant cross-referencing of society and nature that need to be emphasised, not simply the moral and military conclusions drawn. Consider Brigadier-General Sir Reginald Clare Hart's essay of 1911, 'A vindication of war', published in the journal *The Nineteenth Century and After*. Again it is the metaphors – war as individualised wrestling bout, volcano, blood restorer – which are so striking, so worthy of attention and investigation. Hart argues that war is in a real sense a universal vital principle – like the rhythmic rise and fall of the lung, or the beat of the heart, world history is a passage between conflict and peace:

> Just at first [after 1815] all Europe was exhausted and tired of war – the combatants required breathing time between the rounds – but alas! for the peace-party, the stream of human gore soon began to flow again in undiminished volume, just as the quiescent volcano pours forth its periodical flow of fresh lava. Rest follows effort, effort, rest.[11]

War expenditure and general national 'militancy', he declares, should not be played down. There is no need to apologise for a process which is 'erroneously likened to a vampire sucking the very life-blood of the nation. Militancy demands efforts, and all effort demands blood, but, unlike the vampire, it also creates blood. Without effort there would be no life-blood to suck' (p. 227).

Hart views war as essential for ensuring progress and population control. He proposes a kind of hydraulics of war, close to the

10 See Nye, *The Origins of Crowd Psychology*, p. 146.
11 Hart, 'A vindication of war', p. 227.

hydraulics of sex being advanced by pioneering sexologists like
Henry Havelock Ellis in this period:

> If we close the safety-valves of war, the force accumulates, some-
> thing must give way, and there is a fearful and devastating
> explosion, and, with all our good intentions, one single cata-
> strophe may be more awful than the sum total of all the evils
> from which we have spared ourselves in the previous years.
> (p. 230)

War is positively desirable – 'it is not good for a nation to be too
long at peace' since it 'heralds in stagnation, decay and death'
(p. 237). Conflict is vitality and wakefulness, whereas 'Peace for
a nation is like sleep for an individual, it gives time for rest and
recuperation'. But we ought not to sleep too much since war is
'motion and life', and moreover represents a universal law of
nature:

> There has always been constant and deadly war in the vegetable
> as well as in the animal kingdom, indeed, ever since the con-
> ditions of this planet permitted the existence of the lowest forms
> of organic life, and it has only been by war that from these
> humble beginnings it has been possible by evolution and natural
> selection to develop so comparatively perfect a creature as man.
> (p. 238)

Military slaughter is deemed to be the indispensable tool for culling
populations. Hart speaks of the 'relentless war of extermination of
inferior individuals and nations'. Inevitably this process will now be
slower than in the past, because 'natural selection is hindered and
thwarted by civilised man' (*ibid.*).

To take a better-known and more influential Edwardian eugenist
commentator, consider Karl Pearson. In *The Scope and Importance
to the State of the Science of National Eugenics*, he insists that the
army's physical condition is the litmus test of the national well-
being:

> Permanence and dominance in the world passes to and from
> nations even with their rise and fall in mental and bodily fitness.
> No success will attend our attempts to understand past history, to
> cast light on present racial changes, or to predict future develop-
> ment, if we leave out of account the biological factors. Statistics
> as to the prevalence of disease in the army of a defeated nation
> may tell us more than any dissertation on the genius of the

commanders and the cleverness of the statesmen of its victorious foe.[12]

Not all eugenists agreed with Pearson that the state should do away with class struggle so as to be better able to fight the *necessary* war between nations.[13] Pearson argued (with the Boer War very much in mind) that when wars cease mankind will no longer progress, for there will be nothing to check the fertility of inferior stock.[14] As another contributor to the *Eugenics Review* was to spell out the advantages of war (in 1911):

> It may be that an occasional war is of service by reason of the fact that in time of danger the nation attends to the virility of its citizens. Public opinion at such times will not tolerate the perpetuation of feebleness. In any prolonged period of peace soldiers tend to lose their hold on the imagination of the community and to be esteemed as puppets for parade on state occasions.[15]

In such views, the Boer War could be seen as a rude but salutary physical and moral 'awakening' for the nation. Indeed the very crisis over the supposed unfitness of military recruits was said to be efficacious; the image of a strong army was considered crucial, the key symbol of fitness or degeneracy for the nation: 'A military service is, therefore, eugenically useful because it keeps prominently

12 Pearson, *The Scope and Importance to the State of the Science of National Eugenics*, p. 10.

13 But then again Pearson (1857–1936) was no ordinary eugenist. Strikingly, his later evolutionist war views ran very much against the grain of his own middle-class English Quaker background. After mathematics at King's College Cambridge, Pearson had studied law, philosophy and maths in Berlin and Heidelberg. He changed his name from Carl to Karl and wrote *The New Werther* in homage to Goethe. He suffered the orthodox crisis of orthodox faith and wrote a passion play that attacked traditional Christianity before returning to a professorship in maths (increasingly indeed to social statistics about race, disease, hereditary disorders, crime and so forth) and eventually, under Galton's patronage, to a chair of eugenics at University College, London. Earlier influenced by Fichte's view that the interests of the people were best expressed by the state, Pearson would come to see himself in the British context as an iconoclastic rationalist, able to draw German thought, particularly of the left, together with Darwinism into a socialist advocacy of imperialism. He was in his own eyes a member of a caste of experts whose views were above the petty interests and distortions of party politics. Hence *The Ethics of Freethought* (1881) and *The Grammar of Science* (1892). Elected to the Royal Society, Pearson was awarded the Darwin medal in 1898. See Kevles, *In the Name of Eugenics*.

14 See Pearson, *National Life from the Standpoint of Science*.

15 Melville, 'Eugenics and military service', p. 54.

before the community ideals of physical fitness and efficiency as well as of courage and patriotism' (p. 54). The insistence here is that 'historically, through the selection of hardship and struggle and through that of the community's high esteem, warfare has brought men to such a level of virility as they have attained' (*ibid.*).

Others, as we have already seen, were unconvinced. They were able to quote Benjamin Franklin, who had deplored a standing army for removing the 'flower of the nation' from reproduction.[16] According to the American naturalist, David Starr Jordan, doyen of Stanford University (author in his spare time of children's stories, council member of the American Association for International Conciliation, prolific writer from the 1890s to the 1920s on plants, fishes, organic evolution, the survival of the unfit, war and waste), peace was eugenically beneficial. Writing before the outbreak of the First World War, Jordan declared: 'The delusion that war in one generation sharpens the edge of warriorhood in the next generation, has no biological foundation. It is the man who is left who always determines the future' (p. 96). As all of these pro- and anti-war arguments suggest, the appeal to 'the man' rather than the couple who produce offspring is indicative of a pervasive phallocentrism – which is not to say that women were not, elsewhere, the source of powerful eugenist militarist worries.[17] Jordan discusses the Roman Empire as an example of the bio-political cost of military conflict. The best men are sent off to fight wars. Since 'The decline of a people can have but one cause, the decline in the type from which it draws its sires' (p. 98), militarism and war are necessarily dysgenic, a cause of degeneration.

The shared pacifist and militarist assumption of evolutionary naturalism's central relevance to the human activity of war has of course endured beyond the Edwardian period and was very much a shared European and North American dabate. Following on from the American example of Jordan, one might also remark that the socio-biological argument for peace is set out in particularly extreme degenerationist terms in *The Biology of War*, a work translated from the German in 1919, wherein G.F. Nicolai claims that war constitutes a tragic biological degeneration.[18] The fighting

16 See David Starr Jordan, 'War and manhood', p. 95.
17 See Davin, 'Imperialism and motherhood'; Ronsin, *La Grève des ventres*.
18 G.F. Nicolai was a leading heart specialist and professor of physiology at Berlin University. He was incarcerated in a German prison soon after the outbreak of the First World War on account of his strongly expressed anti-militarist views. It was there that he produced his book before escaping to Denmark. *Die Biologie des Krieges* was published in Zurich in 1917.

of the First World War, he warns, ensures that the 'unfittest', the 'physically inferior', survive:

> Children and old men are protected by Government, but besides them the blind, deaf and dumb, idiots, hunchbacks, scrofulous and impotent persons, imbeciles, paralytics, epileptics, dwarfs and abortions – all this human riff-raff and dross need have no anxiety, for no bullets will come hissing against them, and they can stay at home and dress their ulcers while the brave, strong young men are rotting on the battle-field. (p. 82)

These 'stay-at-homes', the 'idiotic and sickly indigenous race' are 'producing the generation to come' (p. 84) with disastrous long-term effects. To be thus aware of war's dysgenic results, he argues, is to lose all romantic illusions about the slaughter. It is true, Nicolai confesses, that war emanates from a deep human desire, stirring us 'to the very depths of our being' and awakening 'primitive and hallowed sentiments which we collectively call patriotism' (p. 225); yet war is 'wrong, harmful and needless' for the healthy nation. Whatever the robustness of the desires embedded in the drive to fight, wars in fact constitute the spasm of the degenerate, 'the last great carouse of which even a degenerate nation can dream' (*ibid.*).

What is so interesting in this material is not so much, or at least not only, the moral viewpoint adopted (advocacy of peace), but the historically shared 'common sense', the evolutionary and degenerationist terms, the socio-biological links, assumed in the text. It would be possible to trace in far more detail than I am able to do here the complex ramifications of the metaphor of war in nineteenth-century biology, anthropology and evolutionary psychiatry; or to analyse more closely the figure of war in Darwin's own work. After all, one has only to turn to the last pages of *The Origin of Species* (1859) to find a figurative conflation of the idea of war and the sublime:

> Thus, from the war of nature, from famine and death, the most exalted object which we are capable of conceiving, namely, the production of the higher animals, directly follows. There is grandeur in this view of life, with its several powers, having been originally breathed into a few forms or into one; and that, whilst this planet has gone cycling on according to the fixed law of

---

Distribution was forbidden in Germany. Translations of Nicolai's work appeared as far afield as Japan (1929). An edition prefaced by Romain Rolland was published in Buenos Aires (1932).

gravity, from so simple a beginning endless forms most beautiful and most wonderful have been, and are being, evolved.[19]

Neither romantic, Hegelian nor social evolutionary idealisations of war during the nineteenth century were the exclusive preserve of any one European power or culture, although the attempt to nail the other side as the exclusive philosophical home of war was to be a consistent feature of First World War propaganda. Whilst it was certainly a commonplace of writing on both sides between 1914 and 1918 to fix the provenance of 'war desire' outside its own bounds, it was Allied propaganda which perhaps made the most sustained intellectual effort to map out the history of the enemy's bellicose philosophy, psychology and history. Commentators have located this Germanic war lineage in the distant past. Whether it began with Charlemagne or only Luther, it came of age with Hegel. The pernicious tradition was then bequeathed to the historian Treitschke, the bloodcurdling General von Bernhardi, or even the 'mad' sneering philosopher, Nietzsche.

Undoubtedly these figures *did* provide a richly gruesome resource for Allied propagandists. For Bernhardi war is synonymous with national virility, peace with impotence. Battle is the theatre for the inevitable biological struggle for survival and mastery – new bloods sweep away the old and feeble as Roman decadence had been overthrown by the invading armies of the North. War, moreover, is the method of achieving not only victory but also identity. As Treitschke insists, 'Again and again it has been proved that it is war which turns a people into a nation'.[20]

For Bernhardi and Treitschke a people represents a stock which develops over generations. History itself is on the side of the strong; indeed history is all about virility. According to Treitschke: 'The features of history are virile, unsuited to sentimental or feminine natures' (vol. I, pp. 20–1). It is stock, race, blood which elevates the nation above a mere legal abstraction: 'In a nation's continuity with bygone generations lies the specific dignity of the State' (*ibid.*). Wars explode the pernicious myth of utilitarianism, declares Treitschke, echoing Hegel, since the individual is prepared to sacrifice supreme self-interest – life – in war. Modern war transcends questions of material calculation, and hence enshrines 'something positively sacred' (vol. I, p. 15).

19 Charles Darwin, *The Origin of Species*, pp. 459–60. For a meticulous exploration of the metaphor of war in the work of Darwin and within the whole tradition of evolutionary naturalism, see La Vergata, *L'equilibrio e la guerra della natura.*
20 Treitschke, *Politics*, vol. I, pp. 20–1.

We have learned to perceive the moral majesty of war through the very processes which to the superficial observer seem brutal and inhuman. The greatness of war is just what at first sight seems to be its horror – that for the sake of their country men will overcome the natural feelings of humanity, that they will slaughter their fellow-men who have done them no injury, nay, whom they perhaps respect as chivalrous foes. Man will not only sacrifice his life, but the natural and justified instincts of his soul; his very self he must offer up for the sake of patriotism; here we have the sublimity of war.... War with all its brutality and sternness, weaves a bond of love between man and man, linking them together to face death, and causing all class distinctions to disappear. (vol. II, pp. 395–6)

Treitschke likens the nation state to a collective personality rather than an organism, since as he puts it 'will is the State's essence' (vol. I, pp. 18–19). The world can never be reduced to one state, he argues, any more than humanity can ever be reduced to one people. To seek to abolish war or conflict, as in some pacifist dream of perpetual harmony and peace, defies instinct. Permanent peace is dismissed as a 'mutilation of human nature'.[21]

For Treitschke war is even a kind of compassion – like euthanasia, it puts the weak and dying out of their misery:

In the unhappy clash between races inspired by fierce mutual enmity, the blood-stained savagery of a quick war of annihilation is more humane, less revolting, than the specious clemency of sloth, which keeps the vanquished in the state of brute beasts while either hardening the hearts of the victors or reducing them to the dull brutality of those they subjugate.[22]

Not only peace but the very desire for permanent peace is taken to symptomatise an ailing, decadent and degenerate age. The war-drive is at once the sign of health and the medicine for degeneracy: 'Most undoubtedly war is the one remedy for an ailing nation.'[23] Prussia is only comprehensible, we are reminded again and again, as the outcome of a long history of fierce war and struggle:

---

21 Treitschke, *Organisation of the Army*, p. 12. *Cf.* 'The grandeur of history lies in the perpetual conflict of nations, and it is simply foolish to desire the suppression of their rivalry' (Treitschke, *Politics*, vol. I, p. 21); 'Without War no State could be' (vol. I, p. 65); or again: 'In war the chaff is winnowed from the wheat' (vol. I, p. 67).
22 Treitschke, *Origins of Prussianism*, p. 55.
23 Treitschke, *Politics*, vol. I, p. 67.

Who can understand the innermost nature of the Prussian people and the Prussian State unless he has familiarised his mind with those pitiless racial conflicts whose vestiges, be we aware of them or not, live on mysteriously in the habits of our people? A spell rises from the ground which was drenched with the noblest German blood in the fight on behalf of the name of Germany and the most sublime gifts of mankind.[24]

But the ideas explored here are as much of a period as of a country, whatever the specific national inflections and differing practical consequences. The 'nation' cannot be simply taken as the pre-given reality which then explains the menace of the texts, since the 'nation' is a concept continually worked on and over, constituted, in the texts themselves. It must also be emphasised that these arguments and images were always contested – they never had a monopoly before 1914. To argue, for instance, that socialist internationalism collapsed and the intellectuals committed treason in 1914 is also to acknowledge the powerful pre-war tradition of opposition to war, militarism, chauvinism and xenophobia. There is a long and never fully extinguished history of cosmopolitan and pacifist thought.[25]

But nor would it be correct to see the war ideas explored in the present study as simply the expression of isolated voices, the work of maverick cranks. For many European commentators they constituted the currency of 'common sense', the taken-for-granted conception of an age of 'victors and vanquished' nations, an internationally shared image of war as purifying crucible. As Zola remarked on the twenty-first anniversary of the Battle at Sedan:

> For a long time it seemed that this was the end of France, that, drained of blood and billions, we would never rise again.... But war is life itself! Nothing exists in nature or is born, grows up or multiplies other than through combat. It is necessary to eat or be eaten that the world may live. Only warrior nations have prospered; a nation dies from the time that it disarms. War is the school of discipline, sacrifice and courage.[26]

Across much of the primary literature discussed above, we find a common commitment to the idea that war and progress are linked, although in De Quincey and Ruskin we evidently also find a romantic recoil from the perceived effects of modern technology. By the 1870s one can detect a powerful and deepening concern and

24 Treitschke, *Teutonic Knights*, p. 19.
25 See, for instance, Brock, *Pacifism in Europe to 1914*.
26 Quoted in Digeon, *La Crise allemande*, p. 278.

hesitation, an increasingly anxious interrogation of the connection between war and progress as also, interconnectedly, of the relation between industrial and national development; as though factories might be the active catalysts of war not simply the passive purveyors of its materials, and war the ruin of modernity not the locomotive of modernisation.[27] Even Herbert Spencer, who had long insisted that evolution, industrialisation, peace and progress were tied together by an inevitable destiny, came to fear a more contradictory and double-edged movement of change and devastation. By the third volume of *The Principles of Sociology* in 1896, faced with the ominous rise of German militarism and the recalcitrant tendencies of British society, Spencer became still more pessimistic. In his conclusion, he foresaw at least the possibility of a catastrophic 'bursting of bonds', a tragic return to war which would constitute 'extinction' not 'progresss'.[28]

---

27 During the 1860s, various commentators continued to express the view that modern industry and war were incompatible. Thus for instance, Émile de Girardin wrote a series of articles in 1868, gathered into one volume in 1870, arguing that the military/warrior phase of history was necessarily succeeded by the advanced, industrial pacific phase, a view which approximates to that of Spencer in England. Heavily influenced by Saint-Simonian ideas, Girardin's vision of war and industry as pulling in opposite directions was to be shattered by 1870–1; see Digeon, *La Crise allemande*, p. 14.

28 See *Principles of Sociology*, vol. III, p. 598, and more generally ch. 24.

9 The Wake of 1870

'THE PRUSSIAN RACE ETHNOLOGICALLY CONSIDERED'

In a paper presented to members of the Anthropological Institute of Great Britain and Ireland on 6 March 1871, J.W. Jackson spoke on the racial aspects of the Franco-Prussian War.[1] He argued that war provided the supreme illustration of the importance of physical and racial differences between national groups. Anthropology demonstrated the decisive place of race in national behaviour, disproving 'philanthropic' arguments about ethnic let alone class or sexual equality:

> Theoretical legislators, like Bentham, together with all those zealous, but rather injudicious philanthropists who deem it necessary to the success of their benevolent undertakings to deny the radical diversity, while they imply if they do not affirm the mental, if not the physical equality of races, are beginning to admit that ethnic specialities are something more than a surface phenomenon, structure being connected with, and in a sense indicative of character.[2]

Modern weapons were an important element to reckon with in the war, but more deeply the conflict seemed to confirm the persistence of ancient racial forces beneath European society. The particular conjuncture of ancient and modern was devastating in its effects:

> And now, as if to confirm us in our views as to the paramount importance of ethnic data, we have the almost pre-historic conflict between Celt and Teuton renewed, not only in all its former force and virulence, but with a certain increase of intensity, due perhaps in part to the scientific appliances and locomotive instrumentalities of modern civilization, which has thus done more to arm the combatants with weapons and provide them with opportunities for mutual destruction than to diminish their

1 J.W. Jackson was a Foundation Fellow of the Anthropological Society of London, former local secretary for Glasgow and Lanarkshire, and a regular contributor to journals such as the *Anthropological Review*.
2 Jackson, 'Racial aspects', p. 31.

ferocity by the culture of those arts, which, according to certain literary authorities, both ancient and modern, are so favourable in the softening of manners. (p. 31)

Jackson argues that the war has proved his own earlier surmises about the character of the Teutons and Celts. The Celts of Gaul, he explains, had been shut in from Tartarian invasions by the Slavons and Teutons, and from Moorish invasion by the Iberians. As a result the Celts:

> present a higher nervous type, and are consequently endowed with more sensibility, susceptibility, and intensity of thought and feeling than their neighbours. This more powerful development of the nervous system as contradistinguished from the osseous and the muscular, constitutes indeed the distinctive characteristic of the Celt; that by which more especially he is separated as a variety from the heavier Teuton and harsher Iberian. (p. 32)

The Celts are capable of brilliance, he claims, but also 'liable to periods of fearful collapse', and so chronically endangered by the 'stronger Teuton'. Might not this anthropological vignette provide 'a key to the history of France, whether in ancient or modern times?' (p. 32–3)

Ancient France, like Britain, 'suffered from the collapse of energy and vigour', hence the Roman invasion; but the Teutonic elements of Gaul in turn led France to a renewed place as the effective imperial centre of the post-Roman world (p. 33). True, the papacy was to become the ecclesiastical successor to the power of ancient Rome, 'But socially and intellectually, if not also politically, the imperial mission, in so far as it has devolved on any one country, has been discharged by France, and she has done this, let us remember, upon the ethnic vigour and renovated racial force obtained through her Teutonic baptism' (p. 34). In spite of 1,200 years of Frankish occupation, the Gauls have always remained the Gauls, a civilised and cultured people, but impulsive, excitable and variable, prone to exhaust their 'vital force' (*ibid.*).[3]

---

3 This is reminiscent of the argument of Gobineau and subsequently of Vacher de Lapouge. According to Gobineau, the infusion of Teutonic blood into much of Gaul had been barely sufficient to found a feudal nobility except in a few areas, like Normandy and some of the Rhenish provinces. Whilst there are thus certain French exceptions – places where 'the great Gothic inundation' has proved truly effective, the Celts on the whole are refined but effete and degenerate; see Gobineau, *Selected Political Writings*, p. 35. The Gothic invasion had for the most part remained a military conquest, not a true racial immigration. Racial defects, we are told, were compounded by the St

The historical destiny of France had thus become a passage from defeat to defeat. The age of Louis XIV was the apogee of French splendour (the French equivalent of Periclean Athens or Augustan Rome), the moment when 'a subdued but gifted people of high nervous temperament having thoroughly re-absorbed their alien conquerors, once more emerged into their appropriate activity and splendour' (p. 36). But there had subsequently been a scientifically discernible slackening off, a gradual dissolution and degeneration which can easily be seen in the passage from the Enlightenment world of Voltaire, D'Alembert, Lavoisier and Cuvier to current nineteenth-century French mediocrity. Jackson contrasts French degeneration to a dawning Teutonic ascendancy. The Teutons 'are as yet but in the morning of their blushing youth, which, however, gives promise of a most heroic and imperial manhood' (p. 39). 'They are the reserve force of the West, which always comes into play when the more nervous race have been exhausted by the morbid excitement of their corrupt civilization. They are the osseous and muscular pole of European humanity' (*ibid.*).

My aim in dwelling on this example at some length is to draw attention to the way in which the Franco-Prussian War came to be anthropologised. For Jackson, Teutonic forces had for centuries literally infused life-blood into the other effete nations of Europe; indeed 'it is impossible to overestimate our obligations to such a race' (*ibid.*). Nevertheless, he is doubtful that Germany will be able to take over the intellectual and cultural mission of France. Germany, he acknowledges, is still politically immature and possesses no great capital like Paris or London. Despite his enthusiasm, Jackson views Germany as ultimately a critical, negative force. Destruction, however, is itself condoned as a necessary cathartic moment in the progress of European history:

> [The Teutons] destroyed the political empire of ancient Rome, but *virtually* they could not erect another in its place. So they shook the Papal Church, but no one will affirm that their conflicting and sectarian Protestantism represents another. And it is

Bartholomew Massacre, the Revocation of the Edict of Nantes and later by the blood-letting of the Terror. The central, unchanging, weak Celtish core was diagnosed as the cause of the defeat of the decadent Second Empire. As Jackson remarks: 'The old civilization, as in Italy, being but imperfectly submerged, soon reappeared; but conversely, the old ethnic effeteness, being also but slenderly supplemented, has again become manifest, and the French are once more Celts, exhausted by an era of empire and civilization, and so awaiting their inevitable baptism of bone and muscle at the hands of their Teutonic, and perhaps, also yet more remotely, their Slavonic neighbours' ('Racial aspects', p. 35).

the same in literature; they criticise and annotate, but they do not *create*, except in music, perhaps. . . . This summed up in other words, implies that the German mind is analytic and not synthetic in its profounder constitution. Hence it can pull down but it cannot build up, or if so, only with enormous labour, as in the achievement of something for which it is imperfectly qualified. (p. 41)

Having thus disposed of France and qualified the potential role of Germany, Jackson is able to bring England into the picture as the saviour of the European march of progress because it demonstrates the desirability of reform in place of revolution, a living monument to the possibility of combining elements of republican democracy and hereditary aristocracy. Jackson concludes that the impetus to future progress lies here:

Regarding England as the geographical terminus of the northwestern march of civilization, I anticipate its culmination on this island, and with this the summation and reproduction of all past imperial missions known to us throughout the historic period. (p. 43)

Jackson's paper gave rise to a considerable and largely appreciative discussion. One contributor caustically noted, however, that the speaker would have done well to draw more of his information on French physique from the scientist Broca and less from the novelist Eugene Sue (p. 44). Another commented that for all the paper's erudite qualities, the fact remained that the French were just as heroic as the Teutons, and so the ethnic explanation of the war's outcome was fraught with difficulties (p. 45). Mr Luke Burke argued that if, as Jackson suggested, England were to take over the mantle of France, '[she] would have to cast off her present parliamentary rulers and their one-sided theories of peace and non-resistance, and take more rational views of the condition of humanity in the present era of the world' (*ibid.*).

The speakers in the discussion share a willingness to take the supposed level or health of each European culture as a sign or symptom of its place in the struggle for mastery. Culture, in short, was at once a weapon and an index of the nation's politico-economic, military and imperial 'destiny'. One member, Colonel Fox, wondered whether the Franco-Prussian conflict had in fact resulted from the ambition of princes rather than the 'war of races' (p. 50). Nevertheless the general presupposition of the very possibility of an international racial, political and cultural competition was largely accepted. In other words, there was to be a kind of

Olympics of intellectual life which in turn might serve as a vital prognosticator of national fates in war and industrial competition.

The *abstract* argument that war determined the destiny of the nation and that the 'best' nation would win, which I have traced and analysed through various instances in the earlier parts of this book, was specifically applied here to the Franco-Prussian War. The participants debated the respective merits of current German and French literature, accepting that such merits or demerits provided the true 'diagnosis' of each culture. For some, Germany remained in 'the rearguard of civilization', its population decidedly 'backward' (p. 48). One speaker defended the evidence for contemporary French racial robustness against Jackson's charge, by countering the claim that there were no real 'master' minds to be found across the Channel in the nineteenth century.[4]

Not all commentators of course shared the same vision of French terminal degeneracy or of German racial deficiency and monstrosity. German political leaders had their supporters, as of course German culture and philosophy had and continued to have its intellectual adherents.[5] Take the following counter-example. In an article entitled 'A few words for Bismarck', published in *Macmillan's Magazine* in 1871, Edwin Goadby (later author of *The England of Shakespeare*, 1881), endeavoured to reject the common view. Bismarck was not in reality the hateful figure of popular legend so often 'regarded as a genius with the will of a monster, the duplicity of a villain, the devotion of a savage, and the lust of a revolutionist'.[6] Goadby reproached other commentators for crushing German history, statecraft and ambition into the single term 'Bismarckism'.[7]

But in this case we should perhaps speak of the displacement rather than the dissolution of the German stereotype. Goadby's strategy in defending Bismarck is not to deny the buried racial evils of the Germanic soul. The aim is to shift the issue from the

4   To prove French genius he cited, amongst other giants, Comte, Broca, Geoffroy
      St Hilaire, Quatrefages, De Candolle, Guizot, Michelet, Taine, Hugo and
      Littré.
         The state of anthropology as a developing science in France and Germany
      was also taken to be an indication of the general racial and political health of
      each country. Dr Blake, for instance, called attention to the eminence of the
      French as anthropologists and anatomists, whilst he considered the German
      school of physiologists to be superior to the French. Bringing a military
      metaphor directly to bear, he argued that 'In anatomy also the Germans *held
      their ground*' (emphasis added) (p. 48).
5   See Wallace, *War and the Image of Germany*.
6   Goadby, 'A few words for Bismarck', p. 339.
7   Goadby's complaint was specifically directed at a recent contribution by
      Frederick Harrison to the *Fortnightly Review*; see *ibid.*, p. 340.

putatively morbid psychology of one leader (which he questions) to the disordered racial anthropology of a people incapable of overcoming their age-old aggressive condition (which he endorses): 'pugnacity is an element in the entire German race, which centuries of settled life and commercial pursuits have been unable to eradicate'. 'The "blood and iron" creed is as old as the race' (p. 341). But it is a doctrine, he argues, also applied in Britain, for instance in policy towards the Fenians. Goadby does not seek to defend Bismarck's gagging of the press, violating of the constitution or general illiberalism, but he 'explains' and 'understands' them by insisting that Germany is both historically and anthropologically unfit for liberalism; it does not have Great Britain's centuries of tolerance behind it. Not only is the Prussian monarchy the creation of war; not only is Prussia always forced to mobilise its military resources, given its vulnerable geographical position (exposed to Russia and France), but Germany itself is the product of war in the 1860s, a dynamic country defeating its 'comatose' neighbours (p. 344). 'War forces events like warm air does plants, and it creates a cohesion which is not producible in any other way' (p. 343).

The Franco-Prussian War produced an extraordinary proliferation of anthropological and medico-biological diagnoses, above all in France.[8] For the important French anthropologist and director of the Museum of Natural History, Jean Louis Armand de Quatrefages, to take another example from the early 1870s, the Franco-Prussian War revealed that Prussia was anthropologically quite alien to Germany, subject to a pitiless mysticism and atavism which stemmed from its 'cross-race', bastardised composition and its distinct 'descent'.

Quatrefages draws on the contemporary speculations of Swedish scholars about the existence of the primitive inhabitants in north-eastern Europe who had preceded the Aryans. Far from being real Germans, the Prussians are identified as Finns or Slavo-Finns. Finns, apparently, never forgive insults whether real or imagined. They avenge themselves at the earliest opportunity and always quite unscrupulously.

The notion of Finnish inferiority had earlier been developed by the Swedish scientist Andreas Retzius, who had founded a whole theory of civilisation on differences in the ratio of maximum width to maximum length of the skull. Skulls on the long side were termed 'dolichocephalic' and represented progressive Bronze Age racial stock; shorter ones were 'brachycephalic', a Stone Age breed. He argued that the 'dolichocephalic' peoples such as the Scandinavians, Germans, English and French were 'endowed with the highest

8   See Digeon, *La Crise allemande.*

faculties of the mind'. 'Brachycephaly' was an indication of a different origin which he described as Turanian, as distinguished from Iranian or Aryan, and which was typical of supposedly retarded peoples such as the Lapps, the Finns or Slavo-Finns, the Bretons and the Basques.

On this basis, Quatrefages linked the Prussians to Finnish primitivism; their current political and military behaviour bore the indelible mark of their ancient savagery. Moreover, 'from an anthropological point of view, Prussia is almost entirely a foreigner to Germany'.[9] The rest of Germany, he argued, had committed an 'anthropological error' in allowing Prussia to dominate it. The error was to imagine that the common language of Germany afforded a common racial-cultural identity.[10]

Quatrefages's theory in the early 1870s aroused some support in France, but heated protests from German writers on anthropology like Adolf Bastian and Rudolf Virchow. Bastian came to the defence of the Finns, those 'poor orphans of anthropology' who had already been maltreated by Gobineau. There were no longer any Finns or Slavs in Prussia, argued Bastian. They had been absorbed or destroyed by the advance of Germanism. 'In the same way as the Indians, they have disappeared, like snow, beneath the rising sun of history.' Or, as he put it in more evolutionist terms: 'The flood of Germans advancing towards the East had implacably rolled back the weaker race following the law of victory to the strongest, the law of the *struggle for existence*.'[11] Prussia, he concluded, 'is a country of warriors; so it was and so it will remain as long as it is necessary to protect the marches of the East, but above all those of the West, against the troublemakers'. German newspaper reports accused Quatrefages of being an 'ignorant naturalist' and a 'learned liar'.[12]

The important German scientist Virchow called his colleagues together to form a commission 'to establish a statistical record of the shapes of skulls throughout Germany, following a method to be formulated by the commission'.[13] As Poliakov observes in his study of the Aryan myth:

9   Quatrefages, *The Prussian Race Ethnologically Considered*, p. 6.
10 See *ibid.*, p. 85.
11 Quoted in Poliakov, *The Aryan Myth*, p. 263.
12 See *ibid.*, pp. 262–3.
13 Their initial intention was to measure the skulls of all soldiers, but the army chiefs refused to collaborate with Virchow and his colleagues. They were obliged to resort to measurements of 'associated' characteristics such as colour of hair, eyes, and complexion. Questionnaires were sent to teachers in Germany, Austria, Switzerland and Belgium requesting the measurements

The investigation assumed gigantic proportions, lasted more than ten years and involved something like fifteen million school-children. Meanwhile the indefatigable Virchow had gone to Finland where he was able to establish that, contrary to the prevalent belief, the overwhelming majority of the Finns was blond. (p. 265)

In 1885 Virchow presented the results of his inquiry to the Prussian Academy of Sciences. It sought to demolish Quatrefages's view. The predominance of blond hair and blue eyes in northern Germany including the territory east of the Elbe showed that the people there were of essentially Germanic stock. In contrast to the Germanic migrations to the west and south – of Goths, Franks and Burgundians who were finally submerged by native populations – those in the east had resulted in a definitive Germanization, 'the formation of a new, purely German, *Volkstum*' (p. 265).

By contrast, the French surgeon and anthropologist Paul Broca, whose subsequent fame rests partly on his work in the identification of the localisation of function in the brain, defended Quatrefages in the 1870s by pointing out that in his view the Prussians had exhibited a peculiarly intense patriotic spirit. This went beyond ordinary love of country as displayed amongst the other German peoples and amounted to a disorder.[14] Nevertheless, Quatrefages was obliged to make a partial retreat. In his lectures in 1872 he declared that his objective had been a purely scientific one, although he admitted he had not been loath to remind the Germans that the Prussians were not their brothers.[15]

This encoding of nationhood in racial anthropological terms, and specifically the attribution of atavism or degeneracy to the German 'race' in general, or the Prussian 'race' in particular, can be followed through to the First World War. Quatrefages's disciple Boule extended the diagnosis in an article of 1914 on war in the journal *L'Anthropologie*. He identified the peculiar German racial strain as, variously, atavistic, degenerate, atrophied and hypertrophied. In addition to its general barbarism, Germany apparently suffered from the conditions of megalomania and gigantism, a true illness afflicting both individuals and the race.[16]

obtained from the examination of all children bar Jews and foreigners; see *ibid.*, p. 265.

14 See *ibid.*, p. 262.
15 See *ibid.*
16 Boule, 'La Guerre', p. 578. For an English example of this kind of diag-nosis, see for instance Merrifield, *Human Evolution in the Direction of Degeneration* [P.].

To take a better known writer, consider the view of that extra-ordinary proselytising crowd theorist and populariser of science, Gustave Le Bon. Germany's behaviour in the First World War, he announced, was the return of the racially repressed which he had predicted for decades: 'All nations possess an aggregate of inherited feelings, which are determinative of their mental orientation.'[17] He noted how 'inextinguishable race-hatreds . . . are among the chief causes of the European war' (p. 30). But it is specifically Germany which displays atavism and an enslavement to its own historical unconscious:

> Men were beginning to forget the dark ages in which the weak were pitilessly crushed. . . . But the belief that the progress of civilization had once and for all destroyed the barbarous customs of primitive periods was a dangerous illusion, for new hordes of savages, whose ancestral ferocity the centuries have not mitigated, even now dream of enslaving the world that they may exploit it. . . . side by side with the intellectual beacons which pilot the scientist in his researches and the philosopher in the discovery of his principles, there exist affective, mystic and collective forces that have no kinship with the intellect, but have each a logic of their own, differing widely from rational logic, which is the foundation of science to be sure, but not of history. These forms of logic which are independent of the intellect were long unknown, for their links are forged in the dim realm of the unconscious, the science of which is but just beginning to be studied. (pp. 19–22)

Beyond the First World War, one can trace still further the deployment and elaboration of such anthropological-degenerationist explanations. Sometimes Prussian 'peculiarity' was transposed into a more generalised conception of German racial anomaly.

In a Foreword to a study which appeared in 1940 entitled *Germany the Aggressor through the Ages* (by F.J.C. Hearnshaw, Emeritus Professor of History at London University), Sir Thomas Holland, Vice-Chancellor of Edinburgh University, raised what he took to be the central issue of Hearnshaw's historical survey: had Germany developed along parallel lines to those accepted charac-teristics which marked the course of civilisation, or had something gone fatally wrong in her racial evolutionary history? He wondered 'whether [German history] indicates any real divergence of a kind which a naturalist would recognise as due to the evolution of a

17   Le Bon, *The Psychology of the Great War*, p. 30.

new sub-species. And if this latter alternative be indicated, when did the branching-off of the civilized stock occur?'[18] The Germans 'represent possibly a new racial sub-species, which makes it physically impossible for them to receive or to emit the truth without distortion' (p. 3).

The moral and political censure of Germany is accomplished here in terms of race, evolution and degeneration. The text remarks upon the 'complete atrophy' of the sense of humour in Germany; Hitlerism is characterised by its 'departures from the normal lines of evolution among civilised communities' (p. 5).

In this long tradition of anthropological diagnosis, the monstrous stereotypes and omnipotent fantasies displayed in German race theory are themselves combated *through forms of racially deterministic theory*. We are told that the 'disease' of Hitlerism might perhaps 'have some limited racial origin, might be localised, and therefore might be dealt with by special political disinfectants or surgery' (*ibid.*). The central disease, we learn again and again, is Prussianism, but this spreads outwards; this insistence on the dissemination and mutation of Prussian elements served to explain the somewhat awkward fact that Hitler was not himself actually Prussian, nor were Nazi supporters so geographically confined:

> If this insanity is of Prussian origin, the disease must have already spread to the brachycephalic races of Southern Germany as well; and the problem before the Allied Governments therefore will not be quite so simple as dealing with an uneducated fanatic like Hitler and the degenerate sycophants who form his 'government'. (p. 6)

TRANSFIGURATIONS

In a suggestive article entitled 'The death and transfiguration of Prussia', the historian Tim Blanning points to the fundamental British historiographical commitment to the idea that something went wrong early in German history (perhaps in the nature of the maurauding hordes who broke up the Roman Empire or in the anti-pluralist tendencies of the German conjuncture of the Enlightenment and Protestantism) which led to a rapid collapse of religion and its replacement by a cult of the state and of duty.[19] This in

18 Hearnshaw, *Germany the Aggressor*, p. 3.
19 For a more detailed analysis of the historiography of 'Prussianism' and the 'Germanic' as the inalienably and immemorially perverse national formation, see Blackbourne and Eley, *The Peculiarities of German History*.

turn is deemed to have combined with an elite-imposed industrial revolution which opened the way for moral relativism, historicism, positivism and social Darwinism – a pathological mix which played itself out remorselessly across modern history. Prussia then made Hitler possible, which was why it had to be abolished by decree of the Allies in 1947.[20] Such was Prussia's fate on 25 February 1947. Law 46 from the Allied Control Commission abolished Prussia with the words 'The Prussian state, which from early days had been the bearer of militarism and reaction in Germany, has *de facto* ceased to exist.'[21]

Whatever the historical origins and historiographical genealogy of this set of perceptions, it is clear that the 1860s is an important watershed in the representation of war and German national character. The short and decisive wars against Denmark (1864), Austria (1866) and France (1870) were militarily swift and culturally shattering. In both 1866 and 1870 it took less than two months for the Prussians to demolish the glorious armies of their opponents. The Franco-Prussian War was taken to demonstrate the arrival of a new and (in the view of many) quite monstrous era of technical sophistication.[22] As the important American economist and social critic Thorstein Veblen put it in a study of Germany in 1915: 'It is only with the new departure of 1870 that Germany has come to take its place in the general apprehension as a singularly striking, not to say unique, instance of exuberant growth.'[23]

Whilst Veblen acknowledges that pre-1870 events had a bearing on this changed perception, 'Anyone who seeks a precise period from which to date this epoch of German history will have difficulty in deciding on any given point earlier than the year named' (p. 59). Veblen accounts for Germany's industrial advance and high efficiency by recourse to natural causes, without drawing on the logic of manifest destiny, providence or national genius. He deploys

20 As Blanning writes: 'Many states have changed their identity, some states have been expunged from the map altogether, but no state – to my knowledge at least – has been formally abolished on the moral grounds that it was a menace to humanity' ('The death and transfiguration of Prussia', p. 442).
21 Quoted in *ibid.*
22 The Franco-Prussian War lasted from 28 January 1870 to the disaster of Sedan on 2 September. On 28 January 1871 the Republican war came to a head with the capitulation of Paris. The war was perceived clearly as a triumph of power and organisation – France outclassed in all departments. In place of the Germany of the artist and the pacifist, or of the jolly and the individualist, contemporaries witnessed a 'new' Germany of collective might (the army), of the automaton, of mechanism, of mathematical precision. *Cf.* Digeon, *La Crise allemande*, p. 59.
23 *Imperial Germany and the Industrial Revolution*, pp. 58–9.

terms like 'race', 'hybrids' and 'reversion' in his account but argues that Germans are not exceptional from a racial point of view; nor are her resources extraordinary. What distinguishes Germany, he suggests, is its *retardation as a state*. The very fact of beginning from behind, 'in an anachronistic state', becomes, paradoxically, an advantage, a kind of springboard enabling a powerful acceleration. Even in the second quarter of the nineteenth century, Veblen remarks, Germany was far behind Britain, indeed the former was stuck 'somewhere in Elizabethan times' (p. 63). Although Germany had great artistic and philosophical achievements to speak of in this period, these 'were not in the line of efficiency that counted materially towards fitness for life under the scheme of things then taking shape in Europe' (p. 62).

Pursuing Veblen's history of Germany's rise a little further, we find that this torpor and placid backwardness began to shift after 1850 when it became possible to discern a bizarre modernising drive: 'But it was not until the second half of the nineteenth century that the alien elements seriously began to *derange* the framework of the archaic scheme' (p. 63; emphasis added). At issue in this diagnosis is a modernity that advances but simultaneously distorts and disturbs; a society that strangely mixes the feudal and the industrial; and in which abnegation becomes duty, insubordination contumacy; in which freedom is equated with obedience (p. 79). In short, Veblen acknowledges, Germany has been Prussianised and not the other way round (p. 81), a fact which constitutes a 'disturbing lodgement in the tissues of the body politic' (p. 64).

The disparity between the economic and the political level of this Germanic modernity is also highlighted by Veblen; modern technology has come to the Germans ready-made without the cultural consequences that its gradual development and continued use has entailed among the peoples whose experiences initiated it, and determined the course of its development. Above all, it is successful war which cements this society:

> Chief of the agencies that have kept the submissive allegiance of the German people to the State intact is, of course, successful warfare, seconded by the disciplinary effects of warlike preparation and indoctrination with warlike arrogance and ambitions. (p. 78)

It is as though a positive lust for war has been inscribed in German psychology. According to Veblen, the German individual has become peculiarly mechanised, at least under the surface. Again we return to this oft-envisaged nexus between war psychology and 'machine mentality': 'In their elements, therefore, the premises and logic of the machine technology are in every man's mind, although they may often be overlaid with a practically impermeable crust of habits of thought of a different and alien sort' (p. 184).

Despite the 'moral qualities' of the Germans, declared the French historian and founder of the *Revue Historique*, Gabriel Monod, in his book *Allemands et Français* (1872), war has revealed or developed a disorder of national character, a field of vice and degeneracy (most of all amongst the higher ranks) which bodes ill for the future.[24] Victory has in itself been corrupting and poisoning to the Germans; demoralisation has been the paradoxical effect of their triumph.[25] German thought, science and religion, he argues, have long tended to idealise force.[26]

According to Monod, French culture and society have also been laid bare by the war. He has much to say on the deficiencies of the army, on the folly of tactics, the weakness of French preparations.[27] But for all that, France is ultimately lovable, where Germany is grotesque, brutal and mean-spirited. Germany, as he noted in a piece on the Franco-Prussian War for *Macmillan's Magazine*, is a contradictory mixture: its people are respecters of women and children; they are capable of nobility and idealism; but they are also prone to rapaciousness, cruelty and merciless greed.[28]

What I am aiming to do here, across these multiple and disparate examples, is to trace out in some detail variants of a powerful shared historical narrative of 'war development' – German in particular – which has governed so much cultural and political thought in Europe from the later nineteenth to the twentieth centuries; a narrative in which machine and human being, industry and death, national allegiance and psychosis are repeatedly, even obsessively, woven together. Sometimes this can be evocative and thought-provoking, but often it is a kind of reflex which precludes insight and responsibility in its frantic national gesture, 'not I'. In this narrative the diagnosis of 'national character' tends to reproduce the reassuring scenario in which 'the enemy' is the exclusive container of evil and danger, the self-enclosed belligerent entity. Moreover, through the externalising location of that disturbing history of machine, death and madness, the non-Germanic national identity

24 Monod, *Allemands et Français*, p. 85.
25 *Ibid.*, p. 89.
26 'Lutherans and Hegelians, pietists and positivists were united in the same adoration of force' (*ibid.*, p. 90).
27 'One must have lived with the soldiers in order to gauge the depth of the trouble' (*ibid.*, p. 114).
28 Monod, 'Souvenirs of the campaign of the Loire', pp. 69–76, 143.

could be clarified and purified. But what so often disturbs this picture is the recognition of a modernising aggressive drive *at large*, with no single national provenance: a 'progress' which risks, or perhaps even delights in, the culmination of war. These analyses of Germany must be seen in the context of a wider European reappraisal of war in the face of the definitive breakdown of the 'Concert of Europe' and the remodelling of armies themselves, which can be consulted elsewhere.[29]

Take then one further example of the cultural diagnosis of 1870, Prussia and the machine age, by the French symbolist poet, mathematician and philosopher, Paul Valéry. His short study of Germany, initially published in an English magazine, the *New Review*, in 1897 and reissued in France as a book in 1924, was entitled *Une Conquête méthodique*. In Valéry's account, a summary of German national character was combined with a wider critique of modern mechanisation at large. The mediocrity of the individual, to be witnessed above all in modern Germany, he argues, is not the mere by-product of industrial-economic 'progress', but its necessary precondition. Modern progress depletes creativity; the great musicians and philosophers are dead for ever. In their place there has emerged an anonymous social machine, dependent on faceless operators whose prime duty is obedience – the military command structure is thus the epitome of the 'ideal' modern socio-economic arrangement.

Germany, Valéry continues, is the emblem of the modern state: rational, painstaking, brutal and expansionist.[30] It has become a perfectly oiled and irresistible weapon of political power and economic self-interest; it is a war machine unable to imagine any other condition of being, inevitably measuring its European rivals in the same terms. In Germany, individualism is perfectly subordinated to the state; human desire is constantly channelled into that wider body. Every activity, every branch of science is geared to the collectivity.[31] The Germans are experts in economic calculation; they market human desires with scientific precision, penetrating to the very heart of each purchaser's fantasies of consumption, analysing the minutiae of existences and desires. In this vision, Germany is viewed as a perfect assimilative and adaptational process. 'There is no new substance for which [Germany] fails to find a use; no science for which it does not discover industrial application' (p. 15).

---

29 See, for instance, Challener, *The French Theory of the Nation in Arms*; Howard, *The Franco-Prussian War*.
30 'Germany has become industrial and commercial as she became military – deliberately' (Valéry, *Une Conquête méthodique*, p. 8).
31 See *ibid.*, p. 12.

Germany underwent a metamorphosis from a politico-military to an economic machine, always with 'perfect preparations, generally sufficient execution – and always with results' (pp. 22–3). Whether in politics, economics or war, the German machine was designed to eliminate chance, to proceed only when success is assured: 'these are the remarkable qualities of the military method, "made in Germany"' (pp. 23–4).

The sense of an arranged historical progress which deranges in its effects, a disturbing destiny unfolding in accordance with a master plan, is conveyed in the very remorselessness of the construction which Valéry uses to describe German advance:

> For example she forms her domain through precise actions of war. Then she imposes on Europe this armed peace which all the other states consider abnormal. Then she puts her industry and commerce on to a war footing. Then she creates her military and merchant marine simultaneously. Then all of a sudden she seeks colonies . . .

> (Par example, elle *fait* son domaine à coups de guerres précises. Puis, elle impose à l'Europe cette paix armée que tous les autres Etats s'imaginent anormale. Puis elle met son industrie et son commerce sur le pied de guerre. Puis elle crée sa marine militaire et marchande simultanément. Puis, elle se cherche tout à coup des colonies . . .) (p. 40)

What is supposedly achieved in Germany is a perfect dovetailing of disparate forces into a unified politico-military machine. Moreover, war is subject to rigorous economic calculation. Far from being a 'chivalrous' contest conducted between equals, Germany will henceforth only declare war when success is mathematically certain (p. 26).[32] In Valéry's view, the German state is a war machine unable to imagine any other condition of being. Hence it inevitably measures its European rivals in the same terms: 'Each nation is then considered as a machine producing military energy.'[33] Thus Germany is a fearsomely effective ensemble, still more frightening in peace than in war (p. 35).

---

32 This chilling image of the German war machine was powerfully deployed in First World War propaganda and had to be rebutted by German apologists: 'One sees that the German army is not, as many say, a tremendous machine, but rather a great, living organism which draws its strength and life-blood from all classes of the whole German folk' (*The Fatherland. Truth about Germany*, p. 16).

33 Valéry, *Une Conquête méthodique*, p. 29.

In a useful and panoramic survey of French perceptions of Germany and the Germanic, Claude Digeon provides numerous examples of how 1870 was cast as a turning point.[34] Digeon traces various anticipations of that 'other' Germany before 1870. He notes how Catholics like Veuillot, for example, identify with Austria and against Prussia during the period of the Second Empire. In this case Prussia was cast as rapacious, expansionist and unscrupulous.[35] But the dominant tone in France remained sympathetic to German nationalism. The great tradition of German philosophy was exalted by French writers like Taine and Renan.[36] Germany was the fine nation of truth, science and ideals, cruelly excluded from the imperial glories of France and England. Prussia was cast as the land of Protestantism, freedom of thought and modernity as against the absolutist clerical Austria which symptomatised for many intellectuals the recalcitrance of the old regime. Napoleon III's tensions with Austria, for instance over Italy, certainly contributed to this sympathy for liberal nationalism which was perceived as a great triumph over archaicism and particularism, the gratifying emergence of something akin to French universalist politics. German and Italian nationalism were seen as articulating the interests of this 'enlightenment'; they were, supposedly, the force of the future, seeking the eclipse of a backward-looking nobility. Nationalism was perceived here as the sweeping away of artificial trade barriers, disparate communications and anachronistic political institutions.[37]

But in the 1870s a bewildering range of explanations for the

34. Digeon, *La Crise allemande de la pensée française*.
35. See *ibid.*, p. 20. Cf. Dumas *père*'s tale, published in 1867, *La Terreur prussienne*. Initially unsuccessful, the book was reprinted in 1872 and translated into English in 1915. The Prussians are shown as pillaging, illiberal, violent, filled with hate for the French and dominated by the rapacity of the Hohenzollerns and Bismarck. Prussia is distinguished from the Germany of free and pacific thought, art, sweetness and light. Dumas deplores the fact that Pallas, goddess of science and wisdom, has been transformed into Minerva. The little kingdoms, like Hanover, appear as nostalgic and out of touch, easy prey for the Prussians; see Digeon, *La Crise allemande*, p. 26. There were other exceptional anticipations of the post-1870 discourse on Germany. Quinet, for instance, had already described the pathology of Prussia before Sadowa. As he wrote on 9 September 1866: 'It is not a question of a crisis of the moment, but of a completely new world situation. It is a quite other race of men that enters upon the scene' (*Lettres d'exil*, 1885−6, quoted in Digeon, *La Crise allemande*, p. 28).

Those isolated figures who had 'seen through' the surface of a benign Germany pre-1870 were subsequently fêted as prophetic.
36. See also Thom, 'Tribes within nations: the ancient Germans and the history of modern France'.
37. See Digeon, *La Crise allemande*, p. 19.

German victory as advanced; from laments for the failure of the scientific spirit in France to the theory of the survival of the fittest; from the language of degeneration and atavism to racial-anthropological suppositions about the relative vitality of different bloods.[38] Germany, it was declared, had fallen like a savage on the helpless prey of France. 'Yes, history will reveal with what voracious gluttony Germany has thrown itself upon this much coveted prey.'[39] The French, it would then be argued after 1914, had come to 'know' the 'real' Germans in 1870, whilst the English had only fully 'discovered' them later. As Kipling has a French woman say in the First World War: ' "Remember, *we* knew the Boche in '70 when *you* did not. We know what he has done in the last year. This is not war. It is against wild beasts that we fight." '[40]

Whilst the material losses of the war were quickly superseded, Prussia continued to haunt French culture; there was indeed, in the words of Digeon's title, a 'German crisis of French thought'; a crisis involving both revulsion from and fascination with the methods of the victor.[41]

So what crystallised from the late 1860s was the image of Prussia as a force transforming not only Germany and its own 'better' qualities, but Europe too, conjuring away liberalism, revealing beneath European civilisation the necessity of power. Yet Prussia, it now often also appeared, was not fundamentally European; the creed it proselytised represented something at once primitive, ferocious and paradoxically advanced, a new kind of developmental contradiction.[42]

38   For a discussion of ideas of scientific, medical and social regeneration in France after 1871, especially as expressed in the *Revue Scientifique*, see Latour, *Les Microbes* (translated as *The Pasteurization of France*), p. 23.
39   Driou, *Le Calvaire de la patrie*, p. 116.
40   Kipling, *The War and A Fleet in Being*, p. 86.
41   As Anatol Rapoport comments: 'France too, learned – this time from her defeat. After the short-lived Paris Commune and consolidating its political power, the French bourgeoisie, like the German, entered on the road to militarization. Defeated by a mass army, France staked her future on a mass army. The goal was *revanche*. About this time Clausewitz was discovered by the French military and was avidly read and quoted by the officers.... The offensive (pictured as an irresistible shock of a massed attack) became the French military dogma. We now know the price paid for this confidence, in the vast French losses in the battles of Verdun and at the river Somme in the First World War. It is difficult for us, who have seen the "martial spirit" replaced by the mechanized juggernaut, to believe that even the famous red pantaloons of the French infantry were not discarded for something more modest and less conspicuous until the First World War was well under way' (Introduction to Clausewitz, *On War*, p. 27).
42   *Cf.* Digeon, *La Crise allemande*, p. 25.

This barbarism in German culture was carefully unearthed in European cultural and anthropological commentary between 1870 and 1918. As the timing of foreign editions makes obvious, the extensive translations of Bernhardi and Treitschke were related to a wider wartime diagnostic interest in German 'philosophy'.[43]

Disturbing pre-war predictions and analyses of German national character were reissued during or beyond the First World War to confirm and reinforce war aims or to demonstrate the prescience of certain writers. Thus for instance a series of Joseph Conrad's pre-war and wartime pamphlets were republished in 1919. Conrad identified Germany as a hungry war machine and a 'powerful and voracious organism':[44]

Indeed [war] has made peace altogether its own, it has modelled it on its own images: a martial, overbearing, warlord sort of peace, with a mailed fist, and turned-moustaches, ringing with the din of grand manoeuvres, eloquent with allusions to glorious feats of arms. (p. 57)

Once again the same elements are in play: the derangement of the relationship between the human and the machine, the appropriation of peace by a war drive necessary to the survival of the German state. Whilst the Napoleonic wars one hundred years earlier were still conducted according to certain relatively gentlemanly codes, 'Mankind has been demoralised since by its own mastery of mechanical appliances'. It has 'become the intoxicated slave of its own detestable ingenuity'.[45] Conrad had sensed before the disaster of war, something terrible happening in Germany, a racial arrogance – but directed against other Europeans rather than (perhaps more justifiably in his view) the non-European. He recoiled from 'that promised land of steel, of chemical dyes, of method, of efficiency; that race planted in the middle of Europe assuming in grotesque

43 See, for instance, Treitschke, *L'Avenir des moyens états du Nord de l'Allemagne* (Paris, 1866); *The Fire-Test of the North-German Confederation* (London, 1870); *The Confessions of Frederick the Great and the Life of Frederick the Great* (London, 1914; New York, 1915); *The Organisation of the Army* (London, 1914); *Germany, France, Russia and Islam* (London, 1915); *History of Germany in the Nineteenth Century* (London, 1915–19); *La Francia dal Primo Impero al 1871* (Bari, 1917); *Politics* (London, 1916); *La Politica* (Bari, 1918). Bernhardi, *Germany and the Next War* (New York, 1911; London, 1912); *La Guerre d'aujourd'hui* (Paris, 1913); *How Germany Makes War* (London, 1914); *Notre Avenir* (Paris, 1915); *La Guerra del Futuro* (Buenos Aires, 1921). A translation of Treitschke's *Origins of Prussianism* appeared in 1942.

44 Conrad, *Autocracy and War* [P.], p. 47.

45 Conrad, *The North Sea on the Eve of War* [P.], p. 20.

vanity the attitude of Europeans amongst effete Asiatics or barbarous niggers ... '.[46]

THE DRIVERLESS TRAIN

It was in the 1860s and 1870s, around the image and the social
reality of the railway, that a powerful new vision of war emerged,
in relation both to Prussian victories and to the American Civil
War. The intersection of the image of war as remorselessly efficient
machine and war as deranged vehicle is aptly caught in the railway
image which takes the reader of Zola's Rougon-Macquart novels
into the Franco-Prussian War: at the end of *La Bête humaine*
(1890), beyond the murders, the triumph and the catastrophe of
the railway in the story, Zola describes a driverless train bearing
intoxicated soldiers to the anarchic disaster of war which he will
then painstakingly portray in *La Débâcle* (1892):

> The engine ran on and on as though lashed to madness by the
> strident sound of her own breath.... With no human hand to
> guide it through the night, it roared on and on, a blind and deaf
> beast let loose amid death and destruction, laden with cannon-
> fodder, these soldiers already silly with fatigue, drunk and
> bawling.[47]

The image of bestial and automaton-like forces in the human
being is intercut with the notion of an animated and willed machine,
'invincible, powerful as a hurricane' (p. 55). Phasie, who has
witnessed the passing of thousands of trains before her house at the
level crossing, understands the human and mechanical violence of
the railway: 'Oh, it's a wonderful invention, you can't deny it....
People travel fast and know more.... But wild beasts are still
wild beasts, and however much they go on inventing still better
machines, there will be wild beasts underneath just the same.'
(p. 56)

46 Conrad, *The Shock of War* [P.], p. 16.
47 *La Bête humaine*, p. 366. Regarding *La Bête humaine*, Zola wrote a note in
    his manuscript: 'Mettre le train plein de soldats gais, insouciants du danger,
    qui chantent des refrains patriotiques. Le train est alors l'image de la France';
    quoted in Digeon, *La Crise allemande*, p. 47; *cf.* the interesting comments on
    the metaphor of the engine and the machine more generally in Balzac's and
    Zola's work in Brook, *Reading for the Plot*, pp. 41–7. Other possible titles
    which Zola appears to have considered for his railway novel included *Les
    Carnassiers, Détruire, Ceux qui tuent, L'Homocide, La Suée du meurtre, Sous
    le progrès, Civilisation, L'Envers du progrès*; see Baroli, 'Le Train dans la
    littérature française', p. 230.

The nexus of machinery, bestiality and war was explored in many different ways. Germany was not always the focus. If the timing of Samuel Butler's *Erewhon* (1872) suggests a connection with the military events of the 1860s and the Franco-Prussian War, the adventure is in fact located on the other side of the world in a strange 'lost' society beyond a daunting mountain range. Butler drew on his experiences of sheep farming in New Zealand, but also upon a more immediately European debate about the machine. He projects a vision of a society where machines have become associated with catastrophic disorder and have finally been banned. The people of Erewhon fear the megalomania of their own inventions and thus proscribe them. Even a watch now evokes massive paranoia. In 'The Book of the Machines' section of the story, we are reminded, as in Zola, of the risk of the driverless machine:

> As yet the machines receive their impression through the agency of man's senses; one travelling machine calls to another in a shrill accent of alarm and the other instantly retires; but it is through the ears of the driver that the voice of the one has acted upon the other. Had there been no driver, the callee would have been deaf to the caller.[48]

In *Erewhon* the relationship between driver and engine turns on the possibility of unforeseen breakdowns and catastrophes.[49] As one of the participants has it in the central Erewhonian debate about whether to destroy all machines, there are numerous unknown and dangerous possibilities:

> 'At first sight it would indeed appear that a vapour-engine cannot help going when set upon a line of rails with the steam up and the machinery in full play; whereas the man whose business it is to drive it can help doing at any moment that he pleases; so that the

48 Butler, *Erewhon*, p. 144. Butler goes on to inquire about a future in which 'hearing will be done by the delicacy of the machine's own construction – when its language shall have been developed from the cry of animals to a speech as intricate as our own' (p. 145).

49 *Cf.* E.M. Forster's pre-First World War story, 'The Machine Stops', a response we are told to 'one of the earlier heavens of H.G. Wells'. Humanity is shown to abdicate responsibility to machinery and gradually to decline into feebleness: 'Quietly and complacently, [humanity] was sinking into decadence, and progress had come to mean the progress of the Machine.' But as the machines wind down, humanity discovers itself incapable of restoring control; Forster's protagonist is forced to watch and endure a catastrophe of humanity's own making: 'And at last the final horror approached – light began to ebb, and she knew that civilization's long day was closing' (*Collected Short Stories*, pp. 148–9, 156).

first has no spontaneity, and is not possessed of any sort of free will, while the second has and is.

'This is true up to a certain point; the driver can stop the engine at any moment that he pleases, but he can only please to do so at certain points which have been fixed for him by others, or in the case of unexpected obstructions which force him to please to do so. His pleasure is not spontaneous; there is an unseen choir of influences around him, which make it impossible for him to act in any other way than one. . . . The only difference [between the driver and the engine] is, that the man is conscious about his wants, and the engine (beyond refusing to work) does not seem to be so; but this is temporary and has been dealt with above.

'Accordingly, the requisite strength being given to the motives that are to drive the driver, there has never, or hardly ever, been an instance of a man stopping his engine through wantonness. But such a case might occur; yes, and it might occur that the engine should break down; but if the train is stopped from some trivial motive it will be found either that the strength of the necessary influences has been miscalculated, or that the man has been miscalculated, in the same way as an engine may break down from an unsuspected flaw; but even in such a case there will have been no spontaneity; the action will have had its true parental causes; spontaneity is only a term for man's ignorance of the gods'. (pp. 156–7)

So often both war and mind are cast as machine-like but also as potentially free-wheeling locomotives irreducible to some Clausewitzean political equation. The image of the driverless train is important. For what is being addressed here is how mechanisation in general and mechanised war in particular transports history elsewhere: how causes and effects are powerfully complicated by the course of conflict. In this scenario, rehearsed again and again between the 1870s and the 1930s, machines are perceived as potentially uncontrollable. Often such war worries converged with anti-democratic sentiments; 'the driverless train' signified the human abdication of responsibility, the crippling weight of degeneracy upon the driver. Rudderless modernity, it appeared, manifested itself in the desperate symptoms of ruling class incapacity, mob hysteria and war. *Germinal* (1885) after all belongs to the same novelistic saga as *La Débâcle*.

To pursue this 'driverless train' image of war and social anarchy to a much later example, consider the view of the Mosleyite military historian Major-General J.F.C. Fuller in his study *War and Western*

Civilization, 1832–1932. Fuller gloomily documents the relationship between war and democracy from the Reform Act (with its 'deep and occult' effects)[50] onwards. Behind 1832 lurks the French Revolution, the origin of the age of 'herd warfare', '*sans culottism*', modern bloodshed:

> armies became more and more the instruments of the people, not only did they grow in size but in ferocity. National armies fight nations, royal armies fight their like, the first obey a mob – always demented, the second a king – generally sane. (p. 27)

Between 1871 and 1914, he argues, Germany replaced France as the 'disturbing factor' in Europe.[51] Moreover 'From this date, the history of Europe becomes the history of the world' (p. 131) and aggression the order of the day. War is injected into the very fibre of European civilisation. Peace in Europe is matched by continuing wars of extermination in the colonies. Fuller diagnoses the problem of modern instability as the consequence of a fatal alliance of democracy, mass psychology and science. Modernity means new methods of communication, not least the radio, so persuasive and corrosive to the mass mind. It has produced a new crisis of hysteria: 'unless the war is short, hysteria will become uncontrollable and animalism rampant' (p. 252). The answer is to ensure that wars are brief (since they cannot be eliminated) by developing sophisticated armaments. War advances mankind but tends if prolonged to derange the average mind – that is the chaos of 'destiny' from which there is no possibility of pacifist escape. Fuller exults in the 'surgical' properties of decisive wars:

> War has been the instrument, the surgical instrument as it were, which has cut the living flesh free from the dead flesh. Without war, a putrefaction would have set in and with it a creeping paralysis embracing in its chilly arms not only society and politics, but science and industry. . . . To-day the Western world is still shell-shocked, and its outlook upon war is blurred by hysteria. . . . Be it remembered, however, that war is an activity which cannot be charmed away. . . . War is a God-appointed instrument to teach wisdom to the foolish and righteousness to the evil-minded. (p. 267)

Universal suffrage is nothing less than 'black magic', which threatens the very bonds of society. Fuller points to the gathering

50 Fuller, *War and Western Civilization*, p. 44.
51 See *ibid.*, p. 131.

crisis of history which has culminated in the freewheeling political terrors of the 1930s. Again we are left with a driverless engine:

> Until these phantoms [of democracy] vanish, until this black magic is smitten low, all I see in the crystal of the future is this: the engine of Western civilization, speeding along the lines of narrow thoughts, roaring towards the Abyss.[52]

WAR'S SATURNALIA

It is in the period from 1870 to 1914 that we can locate not only the build-up of arms towards the First World War, but the circulation and escalation of a cultural critique in which war is at once the symptom and potential ruin of modernity. Whilst the signified enemy was not exclusively German, at least in the English invasion-scare literature which expanded so dramatically after 1870, the presumed danger of invasion as such remained remarkably constant and the Germanic peril was increasingly the dominant motif.

Admittedly, the designation of the Germanic peoples as warlike was not invented in the 1860s or 1870s. Since at least the days of Frederick the Great, an anti-Prussian tradition had developed, for instance in France; but what I have tried to show here is how this was both heightened and reformulated in relation to the Franco-Prussian War.

Germany was shown to have a warlike identity and an ability to exploit technology stealthily but ruthlessly. At the same time, the processes of build-up for war were shown to become maniacally deranged, derailed, uncontrollably moving of their own accord. Bismarckian leaders and ruthless Junkers on the one hand, anarchic machines and intoxicated masses on the other. At issue was a scenario of manipulation and power, but also a vision of anarchy and rampaging technology.

It was widely argued that the American Civil War and the Prussian victories in Europe bore witness to a new relationship between technology and war. Moreover, they had supposedly ushered in a new culture and war psychology. 1870 was cast as the moment when society had to face up to the bellicose realities of modernity. As the eleventh edition of the *Encyclopaedia Britannica* in 1910–11 stated: 'Since 1870 it has been recognised that preparation of the theatre of war is one of the first duties of government.'[53] Since then, the *Encyclopaedia* continued, Europe has been covered

---

52   *Ibid.*, p. 9.
53   *Encyclopaedia Britannica*, 1910–11, vol. XXVIII, p. 306.

with entrenched camps, fortified and strongly garrisoned great arsenals and strategic railways. War henceforth was a matter of movement, depending on good roads and railways; and of supply, destructive firepower and organisation. It was not a blind struggle between mobs, but a conflict of well-organised masses, moving with a view to intelligent cooperation, acting under the impulse of a single will and directed against a definite objective. Modern communications were now deemed to be crucial in the successful pursuit of war: from the telegraph to the telephone, from visual signalling to improved field glasses, from balloons to airships. Whilst Prussia had learnt from the disaster of 1806 that the best brains in the state had to be put at the disposal of the warlord, for the rest of Europe 'it was not until 1866 and 1870 that the preponderating influence of the trained mind was made manifest' (p. 305). Or as Colonel Maude added in his introduction to the translation of Clausewitz which appeared in 1908: 'one after the other, all the Nations of the Continent, taught by such drastic lessons as Königgrätz and Sedan, have accepted that lesson, with the result that today Europe is an armed camp, and *peace is maintained by the equilibrium of forces, and will continue just as long as this equilibrium exists, and no longer.*'[54]

The year 1870 was a complex discursive watershed. For many commentators it not only signalled the arrival of new realities of nationalism, but also the triumph of a debased racial psychology, above all Prussianism. For many it demonstrated a medico-biological process of degeneration in the vanquished French and of pathologically successful development in the victors. Above all, perhaps, at least from the position of an increasingly disorientated British classical *laissez-faire* and individualist ideology, it coalesced with the gathering crisis of belief in the autonomous subject. War, it seemed, shatters the very possibility of individual identity.

In the words of the new liberal theorist, J.A. Hobson, 1870–1 'breaks the spell' whereby France represents Britain's major enemy and commentator.[55] It breaks the spell and fragments the picture. The expansionist plans of Russia absorbed British attention.[56] A shifting kaleidoscope of enemies circulated in fiction and in political analysis – Russian, French, German or even ambiguous half-caste conspirators and spies. Excessive fears of foreign states and

54 *On War*, p. 83.
55 See Hobson, *The German Panic* [P.], p. 15.
56 'The name and certain accessories of "Jingoism" date from the organised propaganda of the later seventies against Russia' (Hobson, *Psychology of Jingoism*, p. 15).

exaggerated patriotic fervour were rife, Hobson complained. To quote the title of his turn-of-the-century study, a peculiar 'psychology of jingoism' had taken root in modern society.

This late nineteenth-century word 'jingoism' is important. Drawing upon an older word 'jingo', meaning a piece of conjurer's gibberish, 'jingoism' was coined in 1878 to describe an extreme and distorted patriotic enthusiasm for Disraeli's use of the British fleet in Turkish waters to resist the advance of Russia. Hobson insisted that processes of nationalism, patriotism and militarism were taking a psychologically warped and even pathological direction. He lamented the perversion of true social allegiances and viewed jingoism as a destructive negative force, an 'introverted patriotism whereby the love of one's own nation is transformed into the hatred of another nation, and the fierce craving to destroy the individual members of that other nation'.[57]

According to Hobson, these processes of jingoism were unprecedented, quite different from some ancient 'war-spirit'. Nor was jingoism exactly primitive; it was indeed 'essentially a product of "civilized" communities' (p. 12). The move to the town and the city, rapid communications of people, goods and ideas, urban conditions, the strained constitutions and the nervous, neurotic disposition of the city dweller was said to produce an acute susceptibility to the morbid excitements of war:

> The neurotic temperament generated by town life seeks natural relief in stormy sensational appeals, and the crowded life of the streets, or other public gatherings, gives the best medium for communicating them. This is the very atmosphere of Jingoism. A coarse patriotism, fed by the wildest rumours and the most violent appeals to hate and the animal lust of blood, passes by quick contagion through the crowded life of cities, and recommends itself everywhere by the satisfaction it affords to sensational cravings. (p. 8)

Hobson drew on a wider context of Victorian evolutionary and degenerationist theory – from Lombrosian criminology and Tylor's anthropology of 'survivals' to Le Bon's crowd psychology; in short he writes in relation to an elaborate range of theoretical models charting the possibility of relapse and primitivism.[58] For Hobson, however, 1870 constituted a key moment when those processes not only erupted but perhaps became dominant.

57 *Ibid.*, p. 1.
58 See Hodgen, *The Doctrine of Survivals*; Nye, *The Origins of Crowd Psychology.*

The whole society, Hobson suggests, has been transformed into a gigantic crowd, with a collective mind.[59] Indeed the most astonishing phenomenon of this 'war-fever' is the 'credulity displayed by the educated classes'. All reason is abandoned in the 'sudden fervour of this strange amalgam of race feeling, animal pugnacity, rapacity, and sporting zest, which they dignify by the name of patriotism' (p. 21). Refined English ladies, quiet businessmen and aged, mild-mannered clergymen, he notes caustically, were all longing to twist the bayonet into the enemy. This 'pulsation of the primitive lust which exults in the downfall and the suffering of an enemy' is 'convincing testimony', he declares in a play on Darwin's title, 'to the descent of man' (p. 31). Although Hobson candidly refers to the primitive lust of the educated, the cause he posits for this descent into bellicosity involves an elitist argument about the vulgarising effects of 'mass' society. Popularisation inevitably means regression: 'The popularization of the power to read had made the press the chief instrument of brutality' (p. 29). The process he locates is circular: the masses produce the war spirit in the newspapers; the press inculcates barbarism in the people (p. 40).

Hobson's distaste for the masses becomes increasingly sharp as the exposition continues. He talks of the 'democratic saturnalia of Ladysmith and Mafeking Days', the 'craving for blood', the 'black slime of [the Jingoist's] malice', and 'these ancient and abandoned stews of savage lust'. His immediate target is the unfolding Boer War and its public celebration despite the 'flagrant breaches of the very canons of "civilized warfare"'. He cites in horror newspaper articles which call for the slaying of the Boers with the same ruthlessness that would be shown to a plague-infected rat (pp. 31–40).

The Psychology of Jingoism refers to a world of war, a world contaminated by the psychological 'black slime' of fanatical masses and cynical imperialist governments, a Boer War world which surely unmasks illusions about the 'purity' of British national character or an individual's ability to think freely, independently and fairly: The Boer War 'thoroughly explodes the old ideal of John Bull as a blunt, frank man who loves a fair fight' (p. 34). War reveals the darker face of Britain, unmasking hidden and disturbing propensities. Hobson's view of the crowd connects with a much wider investigation of mass behaviour in this period. At once a symbol of the shape of things to come (thus 'masses' versus 'classes', 'mass civilisation', 'mass culture' and, eventually, 'mass media') and the locus of an intensifying social psychological and medico-political concern which was frequently linked back to the issue of

59   See Hobson, *Psychology of Jingoism*, pp. 17–19.

war, 'the masses' became a central term in political and social thought.[60]

In his introduction to the 1908 translation of Clausewitz's *On War*, Colonel Maude points to the way modern war theory has been absorbed into the mass mind. He refers to the incubation of Clausewitzean theory in Europe and cites contemporary crowd psychology as proof:

> This estimate of the influence of Clausewitz's sentiments on contemporary thought in Continental Europe may appear exaggerated to those who have not familiarized themselves with M. Gustav de Bon's exposition of the laws governing the formation and conduct of crowds. . . . It is this ceaseless repetition of his fundamental ideas to which one-half of the male population of every Continental Nation has been subjected for two to three years of their lives, which has tuned their minds to vibrate in harmony with his precepts, and those who know and appreciate this fact at its true value have only to strike the necessary chords in order to evoke a response sufficient to overpower any other ethical conception which those who have not organized their forces beforehand can appeal to.[61]

It is as though Clausewitz on war has been absorbed into the unconscious of the crowd. The power of socialism to harness popular thought and feeling, argues Maude, is quite alarming enough, but imagine how much worse when the 'crowd' is mobilised against an external enemy:

> the Statesman who failed to take into account the force of the 'resultant thought wave' of a crowd of some seven million men, all trained to respond to their ruler's call, would be guilty of treachery as grave as one who failed to strike when he knew the Army to be ready for immediate action. (p. 86)

Across the networks of ideas and images traced out in the preceding pages we find certain patterns of narrative explanation: on the one side, a drive to confine the disturbance of 'war psychology' elsewhere (above all to Prussia); on the other, a tendency to view 'mass society' at large as the rootless and hopeless catalyst of war. Suggesting ancient memories and deep 'racial forces', such accounts often stressed the novelty of the conjuncture: an age enslaved by its own 'racial' desires for conquest and war, armed with ever greater industrial destructive potential.

60 See Briggs, 'The language of "mass" and "masses"'.
61 See Clausewitz, *On War*, p. 85. Maude's introduction is included in the Penguin edition which I cite here.

10 Tunnel Visions

Tales of national invasion, accident and insecurity proliferated in the wake of the Franco-Prussian War. The enemies in fiction obviously altered to some extent in relation to the vagaries of international alliances. Indeed until Germany came to exert a virtual monopoly over British war and defence concerns in the years immediately before 1914, it was precisely the uncertainty about where the main enemy lay which exacerbated the sense of cultural drift and national insecurity. The period was chequered with diplomatic crises and manoeuvres culminating in the Franco-Russian alliance of 1894, the German Navy Bill of 1900, the entry of the United Kingdom into alliance with Japan in 1902, French reconciliation with Britain and the beginning of the *Entente* in 1904, the Moroccan crisis of 1905, the arrival of the *Panther* at Agadir in July 1911.

Popular invasion stories swapped one enemy for the other, drawing on an extensive repertoire of hostile images and stereotypes. Some belonged to the 1790s, others had crystallised on the Continent in the aftermath of Prussia's crushing victories over Austria and France. German unification rather than French revolutionary shock-waves was increasingly seen as the central danger to British security and nationhood. As *All the Year Round* warned in 1871:

> Half a million of men, who have trodden down France and threatened England, may pine for fresh conquests. It may suddenly appear necessary for United Germany to win colonies, and a foothold in Central Asia, Persia or India.... They will fly straight at London, the centre of our wealth.[1]

Or as we are told in Erskine Childers's invasion story *The Riddle of the Sands* (1903):

> Here's this huge empire, stretching half over central Europe – an empire growing like wildfire, I believe, in people, and wealth and everything. They've licked the French and the Austrians, and are

---

1 Anon., 'Sieges of London', *All the Year Round*, 1871, p. 497. *Cf.* in the same volume, Anon., 'Foreign Invaders', a sympathetic piece about French refugees in London, victims of 'the bloody Moloch of War' (pp. 133–8).

the greatest military power in Europe. . . . They've got no colonies to speak of, and *must* have them like us.[2]

In this account the British Empire is seen to be weakened by the sheer dispersal of its interests over the whole globe, the rivalry of its army and navy, the inadequacy of its theory of national defence and of its war machine.[3] Germany, on the other hand, is perceived to be in a much less encumbered position; her energies are not 'dissipated' by such vast and distant dependencies whilst her economy, army and navy are robust and expanding. And on top of all that, Childers warns that the German newspapers are ominously prone to 'rancorous Anglophobia' (p. 236).

> [Germany] has a great army (a mere fraction of which would suffice) in a state of high efficiency, but a useless weapon, as against us, unless transported overseas. She has a peculiar genius for organization, not only in elaborating minute detail, but in the grasp of a coherent whole. She knows the art of giving a brain to a machine, of transmitting power to the uttermost cog-wheel, and at the same time of concentrating responsibility in a supreme centre.[4]

In such stories, foreign preparations for war and invasion were shown to be firmly under way. Germany gave 'a brain to a machine', but was also driven along like a quasi-automaton, propelled by the momentum of its own blind power and greed. Whilst such narratives reflected (and sometimes were orchestrated by) military interests, they cannot be read as simply externally 'functional'. The invasion story negotiated a wide range of existing debates and anxieties, mobilising a multiplicity of meanings, images and anticipations in excess of any demonstrable defence need. Moreover, military or naval strategies were themselves founded on a complex mix of fictions, conjectures and narrative possibilities. One cannot interpret the story of invasion as simply the propaganda device of a prior 'hard-nosed' factual strategy. International military perceptions assessed the enemy's potential response in an interminable circle of pessimism and arms escalation.[5]

---

2 Childers, *Riddle of the Sands*, pp. 80–1.
3 Its navy for instance was 'distracted by the multiplicity of its functions in guarding our colossal empire and commerce, and conspicuously lacking a brain, not merely for the smooth control of its own unwieldy mechanism, but for the study of rival aims and systems' (*ibid.*, p. 281).
4 *Ibid.*, p. 280.
5 As Jonathan Steinberg has argued, a pessimism about Germany's future united many of the German naval and army commanders. 'Pessimism could not be

The prospect of other European powers burrowing away at the security and supremacy of this island had a long history, back across and before the nineteenth century. Gillray's series of anti-French prints in 1796 entitled 'Promised Horrors of the French Invasion' graphically suggested the awful executions, desecration, cruelty, corruption and rape which would follow. The French Revolutionary period was to see the crystallisation of reactionary patriotism. Linda Colley has even claimed that reaction effectively came to monopolise the language of patriotism. Certainly conservative icons of 'John Bull' persisted and multiplied.[6]

stilled by the grandiose fleet plans of the Kaiser, for the fleet challenged Britain. The challenge provoked anxiety, and the anxiety rapidly manifested itself in fears of an.*Überfall*' (Steinberg, 'The Copenhagen complex', p. 44).

Germany apparently feared the consequences of its own expansionist hopes. At issue were incongruities and enigmas of national perception which, as Steinberg acknowledges, go beyond the terms of traditional diplomatic or military history.

On 3 February 1905, when Arthur Lee, the civil Lord of the Admiralty declared in a speech to his constituency that 'the Royal Navy would get its blow in first before the other side [Germany] had time even to read in the papers that war had been declared', this confirmed the German dread.

Steinberg describes the intensification of the reciprocal anxieties in the context of the Anglo-Japanese alliance and the Anglo-French *entente*, the appearance of the 'Dreadnought' in 1906 and the Anglo-Russian agreement. Writing in 1909, Count Schlieffen, by then retired as Chief of the General Staff, expressed these worries: 'An endeavour is afoot to bring all these Powers together for a concentrated attack on the Central Powers. At the given moment, the drawbridges are to be let down, the doors are to be opened and the million strong armies let loose, ravaging and destroying. Across the Vosges, the Meuse, the Niemen, the Bug and even the Isonzo and the Tyrolean Alps. The danger seems gigantic' (quoted in Steinberg, 'The Copenhagen complex', p. 40).

6 As new French stereotypes congealed, John Bull was being deployed in new caricatures and with a new stridency. Between 1784 and 1789, we are told, some 15 or so caricatures had been devoted to John Bull; between 1789 and 1800 more than 100 were to appear. Between 1801 and 1810 the John Bull image found its apotheosis in the art market – in the year 1803 alone 71 caricatures were produced. John Bull is not always portrayed as bellicose; sometimes he is the peace-loving figure, protected by the more overtly belligerent Jack Tar or Jack English. But there is consistency in John Bull's Gallophobia, represented in a miscellany of graphic fantasies of English triumph and French humiliation. In one image, for instance, entitled 'A New Map of England and France, the French Invasion, or John Bull Bombarding the Bumboats', Bull is shown dropping his excrement on to the coast of France and the French fleet. His body is represented by the map of Great Britain and his profile is that of George III. See Surel, 'John Bull', p. 11.

The conflict with Revolutionary France was to see the culmination of a longer process across the later eighteenth century in which the language of patriotism was increasingly deployed by the right. In any case 'radical patriotism' up to the 1770s (and the jolt of the American War of Independence)

In the Napoleonic Wars, 'Loyalty to the government, loyalty to the constitution, and resistance to all change were made the hall-marks of the patriot'.[7] For by then the point was to mobilise the patriotism of the potential soldier. Broadsides like *England in Danger!* insisted that class differences and the suffering of the poor were either the consequence of the foreign foe or were insignificant: 'We are not here made up of two sorts – the very rich and the very poor; but property is well circulated amongst all classes of men, from the king to the cottager.'[8]

But how consistent and uniform is the literature of invasion and national caricature? In his study of the genre of the invasion story, I.F. Clarke insists on the endurance of a single tradition across the century:

> Between Gillray in 1796 and Saki's *When William Came* in 1913, there is little real difference. Only a different enemy and different circumstances distinguish the purpose behind Gillray's picture of the French in London from the purpose that shaped Saki's account of the German occupation of London.[9]

Clarke's argument implies a fundamental continuity, although at other points he acknowledges that 1870–1 marked a watershed.[10] A tale like Saki's *When William Came*, I argue, does not simply retell a time-worn tale of invasion. It also speaks to a specific ensemble of late Victorian and Edwardian fears about metropolitan degeneration, German racial vigour and the vitiating and demoralising effects of East European Jewish immigration, which found expression in the Royal Commission on Alien Immigration (1903) and, albeit in a more compromised fashion as far as the extreme scaremongers were concerned, in the Aliens Act itself (1905).

Saki describes not only a German conquest, but the cultural assimilation and colonisation of an England too enfeebled to have anticipated the threat to national sovereignty, and quite unable to offer any effective resistance afterwards. If it retold anything, it was the widely read story 'The Battle of Dorking' in 1871, with its warning of an avoidable catastrophe of invasion stemming from

was itself very often a largely bellicose, xenophobic and chauvinist affair. See Colley, 'Radical patriotism'.

7   Cottrell, 'The Devil on two sticks: Francophobia in 1803', p. 260.
8   Quoted in *ibid.*, p. 262.
9   Clarke, *Voices Prophesying War*, p. 18. Or again, commenting on a play entitled *Descente en Angleterre*, he remarks, 'There is little difference between this fantasy of the Napoleonic period and the mass production of the British and German visions of future wars in the years before 1914' (*ibid.*, p. 13).
10 See *ibid.*, pp. 44, 58–9.

lethargy, amateurism and commercialism: 'We thought we were living in a commercial millennium, which must last for a thousand years at least.'[11] Whilst some of Saki's 'patriots' continue to believe that England is 'still an important racial unit', others are doubtful.[12] Saki portrays a pathetic national indifference to abjection. Some denizens of this crushed land, however, turn out to be quite alien in sentiment and race, beneficiaries rather than victims of the Teutonic invasion. There are good Jews and bad Jews in Saki:

> 'There are even more of them now than there used to be', said Holham. 'I am to a great extent a disliker of Jews myself, but I will be fair to them, and admit that those of them who were in any genuine sense British have remained British and have stuck by us loyally in our misfortune; all honour to them. But the others, the men who by temperament and everything else were far more Teutonic or Polish or Latin than they were British, it was not to be expected that they would be heartbroken because London had suddenly lost its place among the political capitals of the world, and become a cosmopolitan city ... it has taken on some of the aspects of a No-Man's Land, and the Jew, if he likes, may almost consider himself as of the dominant race; at any rate he is ubiquitous. (pp. 710–11)

In Childers's *The Riddle of the Sands*, two young Englishmen sailing in the North Sea gradually unravel the secret of German naval war preparations. But the story also involves English traitors. More fleetingly it evokes images of other racially dubious and inferior figures whom the heroes have to negotiate and resist. According to the viewpoint of the novel's narrator, standards have to be kept up against Jewish rapaciousness and black barbarism.[13] These dangers are interwoven with the treachery and menace of Germany – 'the systematised force which is congenital to the German people' (p. 236). Most horrifyingly of all, a racially ambiguous 'Englishman', masquerading under the name of Dollman, joins forces with Germany. The good Englishman unmasks his dubious compatriot disguised as a German: 'It was something in his looks

---

11 'The Battle of Dorking', p. 571. The tale appeared anonymously in *Blackwood's*, but was in fact written by Lt.-Col. Sir George Tomkyns Chesney. For the story of its huge success, see Clarke, *Voices Prophesying War*, pp. 36, 144–5.

12 Saki, *When William Came*, p. 732.

13 'I have read of men who, when forced by their calling to live for long periods in utter solitude – save for a few black faces – have made it a rule to dress regularly for dinner in order to maintain their self-respect and prevent a relapse into barbarism' (Childers, *Riddle of the Sands*, p. 14).

and manner; you know how different we are from foreigners'
(p. 78). Faces, bodies, formations of the head are all continually
emphasised in the description; the body is inseparable from class,
moral and national delineation within the story.[14]

Childers emphasises class distinctions within London, 'the
imperial city' (p. 23). Carruthers easily inhabits Pall Mall clubs,
whilst Davies, for all his Oxford education, is definitely a cut below,
as both men know: 'I had known him at Oxford – not as one of my
immediate set, but we were a sociable college'. 'He dressed in-
differently, I thought him dull' (pp. 18, 19). Davies fails to get into
the Indian Civil Service, not quite fitting the milieu and its manners.
He goes instead into a solicitor's office.

The two protagonists have to overcome their social differences.
Before comradeship is achieved, we are shown the moral revenge of
that lower middle-class figure. The somewhat effete Carruthers is
discomforted and appalled by the unexpectedly primitive conditions
on Davies's yacht when the two men take the boating holiday
which leads to their adventure and the eventual discovery of
German war aims. The lowly Davies proves to be stronger and
more heroic than Carruthers – 'a being above my plane, of sterner
stuff, wider scope' (p. 70).

Davies is vigilant in the face of German designs. But this is not so
much the consequence of a desire to avert conflict as better to
prepare for it. Indeed he relishes the coming war as a game – '[a]nd
what a splendid game to play!' (p. 120). The story describes a new
cross-class union which arises, phoenix-like, out of all the petty
jealousies and absurdities of English social snobbery. Class conflict
is shown to be disastrous: the real contest lies between national
units. The story celebrates the bonding of these men in the face of a
common danger, a shared recognition of the difference between the
true Englishman of whatever class, and 'the foreigner' abroad or 'in
our midst'. A growing attachment and friendship develops between
the two adventurers. Whilst Carruthers's social connections are
shown to have their uses, he is prone to lethargy. He needs to be
galvanised by the energetic and resourceful Davies, who is not only
passionately pro-Empire but horrified by the dangers of public

14 Thus Carruthers intuits instantly that the villain's daughter, Clara Dollmann,
    is English: 'Two honest English eyes were looking up into mine'. Her German
    is fluent but the 'native constitutional ring was wanting' (p. 169). Dollmann's
    features, by contrast, are evil: the lips and chin a sure sign of his 'defects of
    character', his 'malignant perfidy' and 'base passion' (p. 216). 'The remarkable
    conformation of [his] head giving an impression of intellectual power and
    restless, almost insanely restless energy' (p. 225). His wife displays 'a notable
    lack of breeding'; she was in short 'unmistakably German' (*ibid.*).

apathy. England, he warns, must be vigilant or perish in a Darwinian-imperial world:

> 'I don't blame [the Germans]', said Davies, who, for all his patriotism, had not a particle of radical spleen in his composition.... '*We* can't talk about conquest and grabbing. We've collared a fine share of the world, and they've every right to be jealous. Let them hate us, and say so; it'll teach us to buck up, and that's what really matters.' (p. 98)

The numerous fictions and protests which greeted the renewed late Victorian and Edwardian project for a Channel Tunnel must be situated within this wider literature of invasion. The Tunnel evoked the dread not only of war and conquest, but also more subtly of miscegenation, degeneration, sexual violation and the loss of cultural identity.

Mooted on and off since the days of Napoleon,[15] the cross-Channel link was proposed once again in the 1880s, and was put before Parliament by Sir Edward Watkin in the form of a Private Member's Bill. It provoked widespread consternation, an intense wariness about the undermining of 'the inviolate sea'.[16] Opponents of the Tunnel ridiculed its defenders as pathetic decadents prepared to compromise Britain's 'moat' either for profit or for no better reason than that they suffered from seasickness.[17] (Queen Victoria's initial reaction to news of the project had, apparently, been relief at the thought of a remedy for her seasickness.) Petitions were organised against the Tunnel, most famously the letter which appeared in the influential periodical *The Nineteenth Century* signed by Browning, Tennyson, Huxley, Newman, Spencer, the Archbishop of Canterbury, newspaper and periodical editors, dukes, MPs, generals, clergymen and other dignitaries.[18] Defence was paramount and could never be sufficient. Elaborate and costly precautions, these petitioners warned, would have to be taken which 'might or might not be practically effective when the critical moment arrived'. The crucial task, as another commentator put it, was to 'close up or destroy, upon occasion, the hole which the promoters would bore through our hitherto inviolate frontier'.[19] Given the unpredictability of the future, what could be more dangerous, the author continued, than 'to unisland England and

15 See Bonavia, *The Channel Tunnel Story*.
16 Marquis of Bath *et al.*, 'The proposed Channel Tunnel. A protest', p. 500.
17 See Anon., 'Musings without method', *Blackwood's*, p. 420.
18 Marquis of Bath *et al.*, 'The proposed Channel Tunnel'.
19 Knowles, 'Imperial and national safety', p. 175.

join her soil to the Continent while Europe is seething with unrest and complexities and perplexities' (p. 175).

These powerful warnings were reiterated in invasion stories. Titles included *The Seizure of the Channel Tunnel*, *The Channel Tunnel; or England in Danger*, *Battle of the Channel Tunnel*, *How John Bull Lost London* and *The Surprise of the Channel Tunnel*. In one tale, French soldiers disguised as Englishmen and mere tourists take the sleepy English by surprise in their beds.[20] Commentators bitterly complained that the link with France would make it infinitely more difficult to check and monitor the foreigner. In *The Surprise of the Channel Tunnel*, the reader is told of the arrival of French waiters, bootmakers, milliners, pastrycooks and miscellaneous other tradespersons on the south coast. They are in fact disguised French soldiers who seize the Tunnel, allowing a waiting army to pour through, 'all of which might have been averted, but for the ill-starred and accursed Channel Tunnel project'.[21]

John of Gaunt's speech from *Richard II* was quoted to express the timeless English horror of violating its own sea barrier:

> This Fortresse built by Nature for her selfe,
> Against infection, and the hand of warre:
> This happy breed of men, this little world,
> This precious stone, set in the silver sea,
> Which serves it in the office of a wall,
> Or as a Moate defensive to a house,
> Against the envy of lesse happier Lands . . .[22]

Gladstone, who had opposed the project in 1882 (ostensibly on the grounds that public anxiety made any parliamentary bill unfeasible), vainly defended the project in 1888 whilst conceding the im-

---

20   See 'Grip', *How John Bull Lost London* [P.], p. 38.
21   Forth, *The Surprise of the Channel Tunnel* [P.], p. 22. This fictional perception of the polymorphous foreign invader was to become the fact of newspaper reporting in 1914. Soon after the beginning of the First World War in an article entitled 'The spy peril', *The Times* warned that: 'The type of foreigner engaged in this work is multifarious. An analysis of recent charges in the police-court shows that the occupations of persons who have been arrested on charges of espionage, so far as they have been stated, were: – Hairdressers (3), German naval pensioner, bookkeeper, music-hall artist, German consul, engineer, waiter, pastor of German seamen's mission, subaltern, student, cook, mariner, cabinet-maker, photographer, director of margarine works, director of oil company, professor of languages, and ship's chandler' (*The Times*, 26 August 1914, p. 5).
22   Act II, scene i. See 'Vindex' [G.W. Rusden], *England Crushed* [P.], p. i. See also Beer, 'The Island and the aeroplane', p. 270, on the tendency to quote only selectively from John of Gaunt's speech.

portance of retaining British 'insularity': the Tunnel could be built, he argued, 'without altering in any way our insular character, or insular security'. On the other hand the realisation of the engineering project would 'give us some of the innocent and pacific advantages of a land frontier'.[23]

The anti-Tunnel literature warned of political, military, racial and demographic risks; it alluded, more or less explicitly, to sexual risks of an explosive and invasive kind. A military railway expert was quite wrong when he argued in 1907 that 'The legend of a Tunnel "resembling the crater of an active volcano, hurling forth, without stoppage, torrents of men, horses and guns, directed towards London", surely need not be taken into consideration'.[24] It *did* have to be taken into consideration. The conflict, it seemed to some, would be a national rape and would lead to further miscegenation. Again, the radical MP Charles Bradlaugh, a defender of the engineering project, hit exactly the sensitive nerve when he spoke enthusiastically, and disturbingly, in 1887 of 'intermingling' with the French in a new peaceful Europe: 'Commerce is an eloquent peace preacher; the frequent and more complete intermingling of unarmed peoples begets distaste of war; national prejudices die away under frequent contact.'[25]

But the Channel was a crucial barrier against agitators, armies and anarchists: 'The silver streak,' declared *The Sunday Times* on 16 April 1882:

> is a greater bar to the movements of the Nihilists, Internationalists and Bradlaughites than is generally believed, but with several trains a day between Paris and London, we should have an amount of fraternising between the discontented denizens of the great cities of both countries which would yield very unsatisfactory results on this side of the Channel.[26]

Whilst Germany or Russia might seize the Anglo-French Tunnel, France was a danger too – no reliable business partner. And if a conduit was not agreed, perhaps it would be built anyway, clandestinely.

In Max Pemberton's popular novel, *Pro Patria* (1901), the French are cast as secretly shoving their 'steel tube' of unknown length across the Channel. Their advance guard would open the English end of the Tunnel, and an army would erupt and surge across the

23 Gladstone, *Channel Tunnel* [P.], p. 32.
24 Anon., *The Channel Tunnel by a Military Railway Expert* [P.], p. 5.
25 Bradlaugh, *The Channel Tunnel* [P.], p. 20.
26 Quoted in Barnett and Slater, *The Channel Tunnel*, p. 61.

sleepy, startled countryside.[27] The project is only thwarted at the last minute through the vigilance of the novel's English heroes.

> *That France attempted to build a tunnel under the Channel to England is no longer denied.... We have seen that the more daring capitalists and fanatics of Paris, having compelled the French government to thrust out a tunnel from Calais, sought to open the tunnel here by taking a farmhouse in an Englishman's name.... The vigilance of one man defeated this great scheme; he shut the gate, as he says, in the face of France. But the tube of steel still lies below the sea. No living man, outside the purlieus of the secret, can say how far the tunnel is carried, or where the last tube of it is riveted. It may come even to Dover's cliffs; it may lie many miles from them.*[28]

A range of fears are pulled together in the novel; from horror at the creeping effects of the French Revolution (and its various sequels – 1830, 1848, 1871) to the danger of other powers 'catching up' with Britain's technological, industrial and economic lead, and 'burrowing' under the Empire. But there are further sexual and racial perils. Pemberton's villain, Robert Jeffrey, is 'creole': a racially ambiguous figure from the colonised world:[29] 'In type a creole, whose "colour" you might detect in the thick lips and the angular nails of well-shaped hands. Hair matted and curly; great breadth of shoulders coupled to a long thin neck' (p. 19). It is the treacherous Jeffrey who directs this French plot to invade Britain via the covertly constructed Tunnel. He has spent the previous sixteen years 'learning to become a Frenchman' having earlier been rejected by the English governing classes with their instinctive and well-founded suspicions and reserve: 'The blue veins in his hands reminded me of ancient prejudices – but they were the fruit of his manners, and not of his birth. We had called him "The Panther" at Webb's. No other word could have described him so well' (pp. 47–8).

Jeffrey's resentment is itself perceived as typical of the 'envious' non-white or off-white races:

27 See Figure 4.
28 Pemberton, *Pro Patria*, p. 315.
29 'Creole' is ascribed various meanings in *The Oxford English Dictionary*, including 'Negro born in Brazil, home-born slave, formerly of animals reared at home'. In the West Indies and other parts of America, Mauritius and so forth, a 'person born and naturalized in the country, but of European or of African Negro race: the name having no connotation of colour'. Sometimes Creole implies white, sometimes Negro. Hence of persons 'Born and naturalized in the West Indies, etc., but of European (or negro) descent'.

There was always, I knew, in this man's mind the sore of his colour and of that which he believed to be the due [cause] of it. He had told me, even as a boy, that he hated the 'white man'. No argument could modify the rankling consciousness of an inferiority which his imagination detected. He hated his fellows because they were not as he. And his temperament followed the traditions of his race. Where he could not bully, he fawned. (p. 49)

Captain Hilliard stumbles across Jeffrey's plot, but only after he has been lured into the 'inferno'. He discovers a huge 'tube of steel thrust into the mud' and once inside hears 'the pulsing machine' at work. Strange powers are at play in the story, 'something akin to terror', but also to sexual impropriety: Hilliard wonders 'Why [he] had been seduced to this place' where Jeffrey is able to exert an almost mesmeric fascination over him:

So subtle was the fascination the man exercised upon me, that I began to wonder if he could compel me, after all, to go back to him. His whistle, echoing shrilly in the trees, seemed to strike a discord in my very marrow. I was afraid, and not afraid; excited in thought yet cool in act; desirous of hearing him and escaping him in the same breath. (p. 234)

Having survived various French murder attempts, he doggedly proceeds in his search to solve the mystery that lies 'below that sheen of the waters', obsessively seeking the entrance to Jeffrey's tunnel. Hilliard finally guesses correctly that the French must have taken a plot of land somewhere near Dover. At the secret estate, he encounters a concealed army of labourers and technicians. In one particularly extraordinary conflation of phallic image and feminised French nationhood we are told that the clandestine workers are ready 'to meet that road of steel which, minute by minute, hour by hour, France thrust out beneath the Channel-bed until it should touch the gardens of England and make her mistress of them' (pp. 237–8). Unfortunately Hilliard finds himself once again Jeffrey's prisoner. In the deep 'shadow of the night', the disconsolate prisoner understands the full extent of the French plot which he had so long feared and hitherto so haplessly resisted. 'I was impotent, dumb, caged at an hour of hours when a man would have given all that life had for him to have uttered but a single word to England and the cities' (p. 249).

The novel sets up a compelling image of subterranean penetration. Deep beneath the surface of the water an engine pulses towards England. The mining machine prepares a tunnel, which will then bear a railway. As in the Franco-Prussian War, the train has arrived

4 'The Dream showed me a lonely house . . . and from a great shaft, a silent army emerging'. Illustration by A. Forrestier (from Max Pemberton, *Pro Patria*, 1901).

as the terrible and startling vehicle of invasion. As Hilliard paces his prison cell, he looks into a bleak future where 'day and night, day and night', 'rolling trains . . . steamed below the frothing waters of the Channel to cast out their human freight upon the grassy down' (p. 249).

In Pemberton's novel, as in so much of the contemporary invasion scare literature both in England and the rest of Europe in what was after all the age of Dreyfus, the fear of internal corruption or sabotage matches the fear of external invasion. In such stories the spy is an ambiguous racial figure, not a 'true' subject or citizen but an impostor, seeking radical political objectives, money, anarchy or perhaps something still more sexually bestial and invasive. In the excruciating verse of one popular pamphleteer against the Channel Tunnel:

It is time we grew wise, and opened our eyes
To the Fenian, Russ, Red, and Wild Rad;
For there something must be beyond – *simplicitee* –

That *all Europe can see*, when they're asking with glee –
'Is it true poor John Boule is gone mad?'[30]

Or again:

By the sea girdled round, and a coast iron-bound,
The Almighty has caused us to dwell;
None to make us afraid, unless *self-betrayed*,
Or – *like idiots* – our *Birthright we sell*. (p. 3)

The poet rouses his readers to protest and to oppose the mad
scheme. Failure implies not only that England will 'bore a deep
hole, and *creep under*' to France, but that all sorts of horrid figures
will crawl back in the other direction. The island is beleaguered,
attacked from inside and outside by unscrupulous profiteers,
menacing foreign armies and insidious anarchists:

Rats and Moles to LET IN,
Will we NEVER BEGIN,
While we've sense to know what we're about!
For – as Heaven placed us here,
We can dwell without fear!
And OUR DUTY is clear –
Stern sharp and severe –
All the Rats and Moles to KEEP OUT! (p. 8)

In Le Queux's 1894 invasion story entitled *The Great War in
England in 1897*, the external menace of the French and their
Cossack allies bearing down on the Home Counties is drastically
exacerbated by the bellicose actions of the London urban residuum:

The scene was terrible. The scum of the metropolis had congregated to wage war against their own compatriots whom they
classed among enemies. . . . Dozens of constables were shot dead,
hundreds of Anarchists and Socialists received wounds from
batons. . . . [31]

Bombs explode and mob war breaks out in an England encumbered
by pathetic bureaucracy and hopeless *laissez-faire* liberal policies.
Indeed Le Queux's message is that political weakness provokes the
war.[32] The London rabble aid the Franco-Russian offensive of 1894

30 Anon., *The Channel Tunnel. A True View of It!* [P.], p. 3.
31 Le Queux, *Great War*, p. 45.
32 'It was our policy of *laissez-faire*, a weak Navy and an army bound up with
    red tape, that caused this disastrous invasion of England' (p. 43). Or in his
    later book: 'To be weak is to invite war; to be strong is to prevent it' (Le
    Queux, *The Invasion of 1910*, p. ix).

and the German invasion Le Queux describes in the later tale of 1906. The metropolis is torn apart by this combined internal and external assault: 'The quiet squares of Bloomsbury were, in some cases, great yawning ruins – houses with their fronts torn out revealing the shattered furniture within' (p. 344). The 'foreign' East End swarms west:

> The riff-raff from Whitechapel, those aliens whom we had so long welcomed and pampered in our midst – Russians, Poles, Austrians, Swedes and even Germans – the latter, of course, now declared themselves to be Russians – had swarmed westward in lawless, hungry multitudes.[33]

There was to be a potential liaison, it seemed, between the 'aliens' or 'outcasts' at home and the national enemies abroad; rampaging hordes 'swarm' across the country, a union of countless outcasts and numberless soldiers.

Malthusian fears about the geometric increase of population which would inevitably produce catastrophe were increasingly concentrated by the late nineteenth century on the differential birth rates of classes in the city, and within the working classes on the fecundity of the 'feckless' compared with the 'virtuous', and on the 'foreign' (Irish, Jews and other 'aliens') over and against the 'English'. Certainly the link between population and war was no new theme; the Malthusian terror of an army of starving babies, ravaged by death and disease, persisted across the century in a variety of new inflections. It is a concern, for instance, given voice in Disraeli's *Sybil* (1845). Mere political 'reform', it seems, cannot address or cope with the sheer onslaught of new 'arrivals'. As Gerard explains to Egremont:

> 'You apprehend me? I speak of the annual arrival of more than three hundred thousand strangers in this island. How will you feed them? How will you clothe them? How will you house them. . . .'
>
> ''Tis an awful consideration,' said Egremont, musing. 'Awful,' said Gerard; ''tis the most solemn thing since the deluge. What kingdom can stand against it? Why, go to your history, you're a scholar, and see the fall of the great Roman empire; what was that? Every now and then, there came two or three hundred thousand strangers out of the forests, and crossed the mountains and rivers. They come to us every year, and in greater numbers. What are your invasions of the barbarous nations, your Goths

33 Le Queux, *The Invasion of 1910*, p. 280.

and Visigoths, your Lombards and Huns, to our Population Returns!'[34]

Soon after, Henry Mayhew's well-known journalistic depictions of the swarming but individuated world of London labouring life were to offer the most compelling picture of this modern invasion – as though all the primitive nomadic forces of the ages had congregated in the city. Mayhew's *London Labour and the London Poor* (gathered into a four-volume edition in 1861–2) assumes a constitutional predisposition to war evident amongst certain wandering tribes elsewhere in the world and within certain rootless sectors of the metropolitan population too:

> The nomad then is distinguished from the civilized man by his repugnance to regular and continuous labour – by his want of providence in laying up a store for the future – by his inability to perceive consequences ever so slightly removed from immediate apprehension – by his passion for stupefying herbs and roots, and when possible, for intoxicating fermented liquors – by his extraordinary powers of enduring privation – by his comparative insensibility to pain – . . . by the pleasure he experiences in witnessing the suffering of sentient creatures – *by his delight in warfare* and all perilous sports – by his desire for vengeance – by the looseness of his notions as to property – by the absence of chastity among his women, and his disregard of female honour – and lastly, by his vague sense of religion – his rude idea of a Creator, and utter absence of all appreciation of the mercy of the Divine Spirit.[35]

Every civilised tribe is surrounded by such 'wandering hordes':

> the 'Sonquas' and the 'Fingoes' of this country – paupers, beggars, and outcasts, possessing nothing but what they acquire by depredation from the industrious, provident, and civilized portions of the community; that the heads of these nomades [*sic*] are remarkable for the greater development of the jaws and cheekbones rather than those of the head; – and that they have a secret language of their own – an English '*cuze-cat*' or 'slang' as it is called – for the concealment of their designs: these are points of coincidence so striking that, when placed before the mind, make us marvel that the analogy should have remained thus long unnoticed. (p. 2)

34 Disraeli, *Sybil*, pp. 137–8.
35 *London Labour and the London Poor*, vol. I, p. 2; emphasis added.

These wandering people persist and perpetuate themselves in the midst of the civilisations around them despite the privations and dangers of their life and the efforts of missionaries to civilise them. They remain physically and mentally inferior (perhaps, Mayhew speculates, because the blood in such cases goes to the surface of the body rather than the brain), displaying 'greater development of the animal than of the intellectual or moral nature of man ... high cheek-bones and protruding jaws' (p. 3).

'It is curious,' Mayhew observes, 'that no one has as yet applied the above facts to the explanation of certain anomalies in the present state of society among ourselves.' He proposes to repair this omission by describing the English equivalents of the Sonquas and Fingoes, namely our 'paupers, beggars, and outcasts', who are themselves to be carefully situated within the graded world of vagrants, thieves, showmen, pedlars, harvestermen, prostitutes, beggars, street traders, coachmen, carmen, watermen and sailors of Mayhew's panorama. Each group has its peculiar aspects, but all share certain underlying qualities and defining features – a slang language, lax ideas of property, general improvidence, repugnance to continuous labour, disregard of female honour, love of cruelty, pugnacity and an utter want of religion. Costermongers 'appear to be a distinct race – perhaps, originally of Irish extraction' (p. 6). In any event they are clearly 'primitive': 'The impulsive costermonger, however, approximating more closely to the primitive man, moved solely by the feelings, is as easily humanized by any kindness as he is brutified by any injury' (p. 213).

The anti-Channel Tunnel voices of the later Victorian period rejoined and massively reinforced the anxious aspect of that tradition, warning of a union between foreign races and the newborn disaffected, nomadic and stunted 'aliens in our midst' – the offspring of a class or a race whose loyalties were highly questionable. The debate about this underwater engineering venture thus became a debate about national identity. On the one hand, national difference was perceived as inviolable, solid as granite. On the other hand, to build a link from England to France was to threaten difference, the symbolic separateness of the Anglo-Saxon, the English, the British, the British Isles or the Empire – each of those distinct but overlapping 'imagined communities'[36] – from the foreign. It was to connect the foreign army and the aliens at home in a devastating aggressive alliance.

In 1913, Sartiaux, one of the French engineers involved in the Channel Tunnel project, speculated that Britain's peculiar

36 See Anderson, *Imagined Communities*.

isolationism could not last for ever: 'It does not seem that the dogma of insularity and isolation can continue.'[37] In his view, persistent opposition to the project clearly lay against Britain's national interest. Isolation was unsuitable for either individuals or nations today. Peoples had every interest in becoming acquainted, making comparisons, mingling, or, as the French text literally and more alarmingly puts it, 'penetrating one another' (p. 575).

But it was exactly this reciprocal penetration which so horrified the scheme's opponents. Winston Churchill, who supported the Tunnel and discounted the military fears as unrealistic, pointed out how the navy had previously resisted or discounted other advances, like steam power, the propeller and the submarine. He went on to recall how his father, Lord Randolph Churchill, had opposed the idea by saying that 'the reputation of England has hitherto depended upon her being, as it were, *virgo intacta*'.[38]

In the view of one opponent in *Blackwood's*, England's isolation was the only guarantee that her purity and originality would stay intact:

> To connect France and England with a railway would be to outrage our national sentiment. Great Britain is an island, and from that happy circumstance are derived the strength and narrowness of our character. We are insular in heart and mind, and we should take pride in our insularity. As London is no mean copy of Paris, like other capitals, so an Englishman, for good or evil, is obstinate and apart. And not only is the intense originality of our English literature due to our insularity; to our insularity also we owe our independence and our spirit of enterprise.[39]

Blackwood's was a setting for staunch opposition to the scheme; for the magazine, the very canvassing of the idea of a rail connection revealed, sadly, a wider degeneration, a weakening or softening of national fibre. Only in the age of the crowd, it was argued, could such a proposal seriously be contemplated; only in an era which had already gone a fair way to losing its identity, could the final blow be struck: 'For the sake of the Crowd, then, we are prepared to break with our honourable traditions and to see the Empire rent in pieces' (p. 423).

I do not suggest here that opposition to the Channel Tunnel must, *a priori*, have been wrong, foolish, or groundless; nor of course that the project was technically viable at the time. There may

37 Sartiaux, 'Le Tunnel sous-marin', p. 573.
38 Quoted in Bonavia, *The Channel Tunnel Story*, p. 47.
39 Anon., 'Musings without method', *Blackwood's*, p. 22.

well have been (and conceivably still are) rational reasons to object
to the expense of the Tunnel, to its ecological price, its social
and political 'opportunity cost'. But beyond such an 'argument'
lay something more complex: a process of cultural projection
and introjection, a system of images and fantasies about underground forces, ejected to France like Pemberton's villain Jeffrey, but
hurtling back upon us through the bowels of the earth.

French culture, it should be pointed out, was not without its own
invasion tales, nor its subterranean horror stories, from Hugo's *Les
Misérables* to the sociologist Gabriel Tarde's Wellesian tale of
a future war-ravaged humanity fleeing deep underground.[40] But
neither of these novels has quite the same register of concerns as are
at issue in this chapter. Indeed there do not appear to be exact
French equivalents to these specific English preoccupations.

Clearly Tunnel stories have a wider purchase in the European *fin-
de-siècle* and scientific futurology.[41] But for English objectors the
Channel Tunnel seemed to evoke particularly disproportionate
fears, the unspeakable image of insidious forces gnawing at the
national body. Tunnel opponents glimpsed in the fixed link with
France a dimly lit gallery of treacherous figures and slimy processes
ready to erupt through the smooth surface of national life. What
could be more literally undermining than a tunnel which would put
an end to the geographical reality of the island and all its cherished
vision of the separate, 'island race', the nation which had proudly
resisted Armadas and Napoleons, if not, admittedly, William the
Conqueror?

The key question, the shared problematic, across many of
these texts turns on a ceaselessly threatened but essential island of
Englishness. That elusive and precarious essence had to be shored
up against all the forces that threatened it, without and within. The
very stories that exposed the foreign danger were simultaneously
engaged in a work of representation to produce the sense of
difference – the assured separation – which could then be placed in
doubt. War is shown to threaten the nation, yet it is through
the narration of the coming war that the terms of England's
(endangered) culture can be clarified.

War, as Hegel had earlier explained, is a means of binding the
nation and starkly separating inner and outer, sameness and difference. Hegelian philosophy of war was itself by now frequently

40 Tarde's *Underground Man* [1896], was translated into English in 1905 with a
    preface by H.G. Wells.
41 For a wider survey of tunnels and the nineteenth-century imagination, see
    Williams, *Notes on the Underground*.

interwoven with social Darwinism. The late nineteenth century produced significant new convergences of 'Darwin and Hegel' (the two names were conjoined for instance in the title of an influential text by D.G. Ritchie).[42] But throughout the century denunciations and criticisms of Hegel had persisted: objections not simply to his idealist philosophy but to his apparent apology for the Prussian state.[43] Hegel's posthumous discursive fate in England provides a powerful example of the suspicion of intellectual and cultural invasion. Hegel, it was feared, had polluted England. He was a philosophical force to be monitored, mopped up and sluiced away. As Kirk Willis has shown, the history of the reception of Hegelian thought in Britain ever since the 1820s is also a history of horrified reactions to Germanic contamination.[44] Even a Carlylean admirer of German culture like John Sterling in the 1840s sounded rather equivocal when he referred to German thought 'leaking' into this country.[45] Willis charts the range of nineteenth-century 'Germanophilia' from Coleridge and Carlyle through George Eliot and G.H. Lewes (although he increasingly vehemently turns away from German idealism into the embrace of Comtean positivism) to Green, Bosanquet, Bradley, Caird, McTaggart and Ward; he also explores the continuing critique of, and anxiety about, the pernicious effects of German idealism, culminating in Moore and Russell's philosophical restoration of empiricism. Hegel was cast in First World War propaganda as a kind of monstrous proto-Nietzsche who had sown the seeds for a deranged philosophy of war which had taken root in Germany. Such perceptions were often conflated, the combined constellation then understood as an unproblematically straightforward idealisation of power, hardness, incorporation and mastery. But, ironically, something of the Hegelian vision of war often underpinned the models of the Allied propagandists who denounced him together with much else of German culture and philosophy.

42 Ritchie, *Darwin and Hegel, with Other Philosophical Studies. Cf.* Richter, *The Politics of Conscience. T.H. Green and His Age.*
43 See Willis, 'The introduction and critical reception of Hegelian thought in Britain, 1830–1900', pp. 96–7.
44 Although at the time Hegel was often known only at second hand. According to one commentator, E.B. Pusey, in Oxford of the 1820s, 'only two persons were said to know German' (quoted in *ibid.*, p. 88).
45 'Thought is leaking into this country, – even Strauss sells' (Sterling to Emerson, 28 June 1842, in E. W. Emerson (ed.), *A Correspondence between John Sterling and Ralph Waldo Emerson*, p. 59). But *cf.* 'There are hardly perhaps three Englishmen living with the slightest idea of what Art is – the unity and completeness of the Ideal' (Sterling to Emerson, 20 February 1844, *ibid.*, p. 84).

Those who anxiously and zealously contrasted French or German literature, philosophy and politics to the supposed tranquillity of their own English world, characteristically dreaded the Channel Tunnel as a conduit for all sorts of dangers – secret armies, literary obscenity or revolutionary currents. Even though the Tunnel proponents offered to loop the railway in and out of the French cliffs so that the navy could destroy the link in a single bombardment if hostilities ever broke out, the fears of military attack from underground were never assuaged. How could they be? The issue went far beyond any hard-headed calculations of risk. Of course there was a real arms race, and admirals and generals on occasion wrote prefaces to the sensational scare stories of invasion in order to encourage greater military and naval expenditure; but the fears expressed went beyond 'real' risks or even *Realpolitik*.

Some commentators on the eve of the First World War, not least Sir Arthur Conan Doyle, campaigned for a tunnel as the only way of counteracting the effect of German submarines.[46] Subsequent historians have considered the military pros and cons of the Tunnel for Britain's mobility in the First World War, and have documented the military and political issues at stake in Hitler's mining schemes during the Second World War.[47] My interest here, however, is twofold: first the sense in which the Tunnel controversy was locked into a wider and highly charged late nineteenth-century cultural debate about the function of war, and the consequences of the putative 'degeneracy' of the English state or 'stock'; and secondly the way the Tunnel literature helped *both to produce and to corrode* the conception of purity, the singular identity so alarmingly at risk.

The fear of racial pollution from abroad or physical decline from within was reflected not least in various Edwardian inquiries and legislative measures. An Inter-Departmental Committee on Physical Deterioration was held in the wake of the Boer War to examine the question of whether potential army recruits were feebler than in the past. It was able to offer some reassurance on this point, evidence that the problems of physical decrepitude had been overblown by the press. Nevertheless numerous commentators continued to clamour for action. On the foreign front, action was under way. An Aliens Act in 1905 was designed to tighten up the borders or, to be more precise, to control the racially and culturally corrupting influx of East European Jews, those supposedly haggling, 'dirty and

46 See Eby, *The Road to Armageddon*, pp. 190–1.
47 See Barnett and Slater, *The Channel Tunnel*; Farquharson, 'After Sealion: A German Channel Tunnel?'

bejewelled Hebrew[s] in the East End' fleetingly but significantly described by Childers in *The Riddle of the Sands*.[48] Whilst some still argued that England was England just because it opened the door to refugees and exiles, others insisted that such largesse could not go on. As the virulently xenophobic Arnold White had said in 1886, England could not continue to act as the world's 'rubbish heap', the sink into which other countries poured their scum.[49]

As well as the inquiry into national deterioration after Britain's ignominious military performance in the Boer War, an Aliens Act, a Prevention of Crimes Act and a Mental Deficiency Act all followed from this concern with boundaries and bodies, purity and danger. Infiltration, corruption and invasion had to be stopped. Proposals were canvassed, sometimes resisted and sometimes enacted, to eliminate enemies without and within – proposals to locate and shore up the welfare of that continually elusive essence called Englishness. The reviled engineering proposals of the Victorian period threatened not only putatively 'real' military exigencies, but a particular symbolic system; yet that system needed the threat in order to be defined; or rather, since it could not easily be defined (there were so many possible definitions), to be delimited. Often this mirage was of an organic community united behind the walls and moats of an idealised feudal castle, or a lovely undesecrated garden. The Tunnel's history raises the difficult question of how far English society in that period needed to grasp at insidious forces and burrowing enemies, from internal outcasts to invading aliens, in order to provide the foundation and the support for a viable mythology of national identity.

---

48 *The Riddle of the Sands*, p. 22. *Cf.* 'I was in the Jewish quarter, striking bargains' (p. 242). On pre- and post-1914–18 European debates about the supposed physical and moral unfitness of Jews for military service (and hence for true national membership), see Gilman, *The Jew's Body*, pp. 42–9.

49 White, *Problems of a Great City*, p. 144.

11 1914: The 'Deep Sources'

the sources of the war reach deep down into the centuries.[1]

One of the characteristics of this war is the perpetual discussion of causes and responsibility.[2]

In an influential study of national character, published in 1927, the former principal of King's College, London, Ernest Barker, rejected the idea of an eternal national destiny.[3] Nations were made and remade; they were a composite of factors and forces, not a single essence. We must beware of 'facile generalizations about immutable national traits' (p. 8). And yet, crucially, each nation has its deep sources, 'profound and abiding permanences in a nation's character; and the heaving of the surface must not blind us to the stillness of the depths' (*ibid.*). English puritanism, for instance, was an historical phenomenon, but it connected with a deeper stratum of 'settled and reflective melancholy in much of Anglo-Saxon literature' (*ibid.*). History and literature are perceived as inextricable and absolutely fundamental in this account, especially since Barker rejects the idea of race itself as the key to national continuity: 'The soil of each country has been washed over again and again by different human species, which have left their representatives in its living population' (pp. 10–11).

The language of racial essences and destinies is here anathematised as essentially foreign, and more specifically Germanic: 'The sovereign people which professes to be a race as well as a people may well be a dangerous monster of centralizing tyranny'. By contrast, the English way, Barker's way, is to insist on a great unifying cultural tradition, 'not the physical fact of one blood, but the mental fact of one tradition'. He does not reject the idea of there being such a thing as the pure kernel of a nation (although it is always then enclosed by other layers), but he partly relocates it

---

1 *The Fatherland: Viereck–Chesterton Debate* [P.], p. 26.
2 Morello, *L'Aggressione della Germania* [P.], p. 5.
3 'Character is not a destiny to each nation. ... We cannot therefore draw up an indictment against a whole nation as eternally cursed, or sing paeans in its praise as eternally blessed, by the destiny of an inevitable character suspended above it for ever' (Barker, *National Character and the Factors in its Formation*, p. 7).

from the body to the mind of the community: 'A nation remains in its essence a fund of common thoughts and common sentiments, acquired by historic effort' (pp. 11–12).

Barker surveys sceptically the factors often held to underpin such common allegiance. 'Protean terms' like race will not do; nor does religion provide a sufficient explanation. He recognises geography's importance – 'Reinforcing the melancholy of tradition, there is the melancholy of the landscapes in which [the Celts] found refuge' (p. 33) – but this still is not a sufficient factor. Nor are these common national thoughts simply a function of language (as the example of Switzerland showed), even if there is usually an extremely close relationship between the history of words and places. In the end, Barker concludes, the nation emerges out of an ensemble of factors.

Having rejected the idea of pure nation races, Barker returns to the view that the particular 'impure' *mélange* is decisive. There are thus good and bad cross-breeds of races. Barker does not reject the idea of pure race as such, but he rejects the idea that any nation actually embodies such purity. British civilisation apparently shares with ancient Greece a benign mix of Nordic and Mediterranean stock. This is 'one of the most fruitful blends'. Conversely:

> There are, it is true, elements which it is better not to mix, because they are so unlike that their offspring, with its ill-assorted mixture of discrepant qualities, will be ill-balanced and un-harmonious. Miscegenation of East and West, or of white and black, has its perils. But interbreeding of the different varieties or races of Europe, which unite a fund of similarity to all their differences, is an entirely different matter. (p. 39)

The history of the British Isles is a history of blending between specific racial qualities – between, say, the 'heavier and more Flemish qualities of the Anglo-Saxons' and the 'lighter but finer temper of the Scandinavian invaders'. But it is also a saga of survival, and sometimes of the unhappy opposition of different and unblended stocks. 'This is particularly evident in Irish history' (pp. 40–41). Race is not destiny for Barker, but it is right (indeed a 'civic duty') to analyse one's blend as a nation. The present is constituted by the past, fashioned by tradition, custom and the 'raw materials' of racial stock. Some blending is good, but too much immigration threatens to engulf the central components of the culture, 'the solid core which is the basis of its tradition. And the nation which loses its tradition has lost its very self.' Culture and tradition have a central place in Barker's attempted reorientation of the terms of nationalism.

The idea of English literature and the English language as fundamental components of nationhood, and of literary criticism as a distinctly national cultural project, even a surrogate social religion, had developed, notably via Matthew Arnold, in the later nineteenth century. It was to crystallise in the context of the First World War, not least in the form of the propaganda pamphlet, a genre to which Barker, like so many other academics, was to contribute prodigiously. Here the terms of the English tradition were to be located through the unrelenting delineation of the non-English.

Consider the war speeches of the literary critic and Oxford professor, Sir Walter Raleigh. 'We shall never understand the Germans,' nor they us, he explains.[4] This incomprehension stems not simply from their devotion to the state, their vulgarity and worship of the mechanical (all anathema to the English) but something deeper: 'the convulsions of war have thrown up things that are deeper than these, primaeval things, which, until recently, civilization was believed to have destroyed' (p. 9). What the convulsions of war actually disclose is never quite made clear, but the sense of darkness and terror cannot be doubted – the speeches refer to Germany repeating 'as if in a bad dream, the old boastful appeals to military glory'; to an imitative Germany which 'haunt[s] me like a nightmare', to a 'Germanized world [that] would be a nightmare' (pp. 12, 27). The conflict is 'a war of ideas' in which it is crucial that we understand the 'ultimate and essential' differences, however difficult they are to bring forth; moreover that we grasp Germany as both precisely controlled and out of control (the configuration explored at some length in my discussion above):

> The German nation is a carefully built, smooth-running machine, with powerful engines. It has only one fault – that any fool can drive it; and seeing that the governing class in Germany is obstinate and unimaginative, there is no lack of drivers to pilot it to disaster. (p. 43)

The attempt to develop a distinctive English literary critical approach was founded on just such a recognition of the abyss between German and English methods. Basil Willey, eventually to take up the King Edward VII Chair at Cambridge, would remark later about the early days of Cambridge English, that:

> the most significant thing, the genuine revolution, was the War of Independence whereby English became an autonomous discipline, free from all alien tyrannies and ancient prejudices. . . .

4 *England and the War*, p. 9.

And so it was with Cambridge English in that decade; fighting for recognition against powerful vested interests, it believed in its own cause and presented a united front to all its critics.[5]

Above all it was the influential Cambridge professor, Sir Arthur Quiller-Couch, who was to delineate English literature in relation to an abhorred German *Kultur*. Writing itself, he declared in 1917, must be sharpened, thereby 'keeping bright the noble weapon of English, testing its poise and edge'.[6] He recognised that English must be understood in a European context, but nevertheless insisted on its radical separateness. English was a war resource, 'a grand patrimony' to carry us through; a language to be husbanded for 'our sons, now fighting in France' (p. 114). Or to quote Willey again: Quiller-Couch 'knew no Anglo-Saxon, hated pedantry and the Germanic type of scholarship'; 'all things Teutonic were at a discount, and "Q" was able to score points off philology, off all accumulations of mere *facts about* literature, and all pseudo-philosophical classifications of literary types and epochs by calling them "Germanic"'.[7]

In its initial period the English School at Cambridge was centrally concerned with such questions of national difference and German perversity, although there were also of course key figures in the early history of Cambridge English, like C.K. Ogden (later to be co-author with I.A. Richards of various seminal studies) who wrote extensively against the war.[8] Cambridge English was firmly established on an independent basis in 1917,[9] but the vexed question of the methods to be deployed in the newly formalised structure of teaching literature was only really resolved in the later 1920s at Cambridge with Richards's vastly influential theory of 'practical criticism'.[10] The 'English method' was to involve a quasi-scientific empiricism and an utter rejection of all mysticism and philosophical idealism – tendencies associated with Germany. At the same time, opposition to the establishment of an English Tripos at Cambridge was itself initially galvanised by anti-German feeling. It did not help that two of the leading reformers in the university, Hermann Breul

5 *Cambridge and Other Memories*, p. 12. *Cf.* 'students and lovers of our literature could pursue their chosen subject without having to endure the alien yoke of Teutonic philology' (p. 14).
6 *On the Art of Reading*, p. 110.
7 *Cambridge and Other Memories*, pp. 13, 17.
8 See Florence, Marshall and Ogden, *Militarism versus Feminism*.
9 See Baldick's lucid account in *The Social Mission of English Criticism*, p. 80.
10 See Mulhern, *The Moment of 'Scrutiny'*.

(Professor of German) and E.G.W. Braunholtz (Reader in Romance Languages) were German-born.[11]

The First World War implicated university teachers, novelists, scientists and journalists far more directly than hitherto in a war of words, literatures and languages, a written appeal to the 'mind' of an army literate as never before, and more broadly to the nation and 'English culture'. The newly systematised propaganda machine took the form of a War Propaganda Bureau under the Cabinet Minister C.F.G. Masterman. It was to include its own Literature and Art Department, drawing heavily on the popularity of novelists like Wells and Bennett.[12]

And yet the question of war language and the widespread intellectual identification with the military conflict cannot easily be explained away as the effect of a narrowly conceived 'propaganda'. My point here is not simply that the terms of the justification for war and of the location of the Prussian 'other' evidently ran back much earlier; but also that wartime propaganda was itself *active* and often unsettled, a continuing and sometimes uneasy attempt to grasp and define national character. First World War propaganda drew on earlier language (tales, fears, images, stereotypes from before 1914, some of which might also be termed propaganda). But it should not be seen as some static rhetoric, proclaiming a message precisely worked out in advance. If it implied and appealed to readers of a particular kind, it also sometimes struggled to define them. However repetitive and declamatory, the flood of pamphlets and images which followed August 1914 was also a crucial process of definition, an ideological work in progress, specifying the aims as the war proceeded, endeavouring to conjure up a sense of nation and to draw the physiognomy of its enemy. In part, of course, propaganda is functional: it is easier and more urgent for the soldier to kill a dehumanised monstrous spectre than a human being. But it also has a wider cultural and psychological purchase, more complex functions of location, reassurance and denigration. My concern is not to seek to 'authenticate' propaganda claims by recourse to documented instances of cruelty perpetrated in the war, nor to

11 See Wallace, *War and the Image of Germany*, p. 165. For a range of wartime repudiations and some defences of Germanic influences on English culture, see Hynes, *A War Imagined*, pp. 68–75. Hynes discusses a variety of academic debates about German culture and philosophy, and records for instance how concerts featuring German music (especially modern music) were cancelled in London in 1914; see *ibid.*, pp. 75–7.

12 Bennett was indeed to become Director of British propaganda in France in 1918; see Baldick, *Social Mission*, p. 87; Sanders and Taylor, *British Propaganda during the First World War*.

verify the actual degree of atrocities committed on either side, tasks which have already been widely undertaken elsewhere in the secondary literature.[13] Nor still do I wish to weigh the pros and cons of historical debates, particularly around the work of Fritz Fischer, about the historical continuity of aggressive German war aims. Historians tell us that the Prussians were the most feared contingent of troops faced by the Allies on the Somme; my intention here is not to seek to prove whether they were or were not more brutal on the field than other groups (for whatever reasons), but to provide a wider picture of perceptions of the Prussian machine from before the war which interconnects with wartime views and experiences.[14] My purpose is to explore the continuities in and ruptures of a language of war through which atrocities and violence, whether real or fabricated, were negotiated and explained.

It is not necessary to rehearse any further the story of the intellectual euphoria that prevailed in 1914, in which war appeared to many commentators, including many artists of the avant-garde, a source of transcendence – the overcoming of narrow materialism, utilitarianism and atomism in modern society. National allegiances across Europe were to result in the dissolution of internationalist sentiments, most notoriously amongst socialists.[15] The First World War was justified and rationalised as a fundamental conflict, an historically and anthropologically inevitable struggle between two inalienably different forces. Sometimes such a view is seen as predominantly German, but there are no shortage of English examples. Beyond the apparently 'epiphenomenal' factors – the 'fuse' of a Sarajevo – lay a separate realm of 'deep structural causes'. The assassination was merely an excuse, 'political and secondary reasons', as J.W. Carliol put it in a 1914 essay entitled 'The inner meaning of the war':

> Underneath, and at the root of this Titanic conflict, antagonistic principles and powers, irreconcilable ideas and ideals, the ideals of faith and the ideals of force are contending. These are the sap of the contention: the very breath of its nostrils and the source of

13 There is a large literature on atrocity propaganda which does seek to explore the historical veracity of its claims. See for instance an early study, Read, *Atrocity Propaganda*; or recently, Williams, ' "Remember the Llandovery Castle" '.

14 On British troops' differentiation between the various German units facing them, see Middlebrook, *The First Day on the Somme*, p. 58. 'The Prussians were recognized as the most aggressive of their opponents, the Silesians were reputed to be lazy and the Bavarians fairly easy-going.'

15 See Stromberg, *Redemption by War*; cf. the interesting recent discussion and range of examples in Timms, *Karl Kraus*.

its vigour. But for them this war, with its world-encompassing issues, would never have come into being; and until one of them has been utterly vanquished it cannot reach its end.... It is this great fact – the fact that the conflict is a conflict of spirits – which distinguishes the present war from all the wars preceding it.[16]

It was a commonplace to argue that the First World War could only be grasped by probing deeply into race, psychology and biology. 'To understand truly the psychology of the nations which have plunged headlong into the present unprecedented war, it is incumbent not only to study their elaborate historical records, but to take a deeper biological account of the native vigour of each national stock,' declared the psychiatrist Henry Maudsley in an article written in 1916, but not published until 1919 due to objections from the censor.[17] It was Maudsley's reference to 'the present murderous war', to Christian rationalisations of the conflict on both sides, to Britain's own 'subjugation of weaker and so-called inferior races' and to the absurdity of belief in a post-war utopia of permanent peace which no doubt caused the problems with the censor, even if he also referred to the Englishman's 'inveterate national prejudice in favour of fair play' and to 'brutal Prussian nature', 'bestial defilements' and 'their gospel of terrorism' (pp. 75–85).

Despite all the late nineteenth-century philosophical rationales, scientific justifications and institutional mobilisations for war, official propaganda between 1914 and 1918 characteristically sought to disavow all bellicosity. The question of the origins of war 'desire' was a continual issue in the more general labour to fix the meaning and significance of the conflict.[18] From across the political

16 Carliol, 'The inner meaning of the war', p. 730.
17 Maudsley, 'War psychology: English and German'.
18 Famous authorities on both sides lent their support to inquests on the causes of the war and the real desires of the participants. Durkheim, for instance, participated in an inquest into the key issue of which side was responsible for the original wish: 'Which nation wished for war in preference to peace and what was its reason for preferring it?' The answer given: Austria and Germany (see Durkheim and Denis, *Who Wanted War?* [P.]). To take another example, consider the following from 1915: 'the Germans have the audacity to assert that they have not desired the war, and that they have been driven to it by us. We, I repeat, hold it to be proved with absolute certainty that, in the month of August last they were the responsible authors of it.... Only ask yourselves if ever there was a people, in like degree as the German people,... prepared for war as an essential and natural function of their national life; consider how many active and formidable motives were heaped one upon another: material interests, the thirst for gold, a natural barbarity of character, patriotism morbidly stimulated by folly of arrogance, a complex

spectrum, commentators sought to locate the provenance of the fundamental emotions and determinants. In answer to the question 'What is the war about?', a socialist pamphlet entitled *Shall We Go On?* (1918) confessed: 'Though, after nearly four years of war, this question should be ridiculous, it is in truth not so; for the real object of the war has only slowly manifested itself, and is only now coming into clear visibility.'[19]

The war was cast as a complex process of *becoming*: not until its 'dusk', as it were, had the war's own inner dynamic become manifest. The pamphlet went on to argue that only with the 'pacifist revolution', namely the Russian Revolution, had the 'positive and original object' of Prussia became apparent. With the benefit of hindsight – the perspective which opened beyond the Russian Revolution – the original German drive to war could be fully understood: 'it is to obtain an unfettered freedom to exploit the Slav peoples for the purpose, first, of dominating Europe and, afterwards, of dominating the world. This having now been proved to be the aboriginal purpose of the war' (p. 1).

Whilst the initial causes could be recounted, the course of the war brought forth new deeper motives for fighting, a range of atrocities which revealed with hindsight the truth of the character of the enemy. Truth, like victory, emerged slowly, even imperceptibly, in the slow war of attrition. This was a source of some anxiety. Propagandists often confessed their worry that the war aims had become too complex, prone to blur, or at least to confuse the ill-informed: 'Even after more than a year of war,' declared the Earl of Cromer in a pamphlet in 1916, 'there are still visible symptoms, which seem to indicate that some sections of the community are not as yet fully alive to the importance of the issues at stake in the present contest.'[20] The great danger, he then revealed, was the phenomenon of Pan-Germanism, that chauvinism which slowly spread 'tentacles' across the globe, whilst appealing to the 'false code of civilization termed German *Kultur*' (pp. 14, 16).

Europe was 'faced by a phenomenon of mentality hitherto unknown in the Western World', namely 'the Prussian cult of war'.[21] Germany had become a war machine, a monstrous figure, '[w]ith Kultur in one hand and a bomb in the other'. Whilst it was admitted that British imperialism was 'still a source of injustice here and

---

and powerful mysticism' (Lavisse and Andler, *German Theory and Practice of War*, p. 48).

19 Anon., *Shall We Go On? A Socialist's Answer* [P.], p. 1.
20 Earl of Cromer, *Pan-Germanism* [P.], p. 3.
21 Anon., *Shall We Go On? A Socialist's Answer* [P.], p. 3.

5   'Gott Mit Uns'; or German Kultur.
Three cartoons drawn by Hugh
Thomson for the pamphlet by Sir
Isidore Spielmann, *Germany's Im-
pending Doom*, 1918.

there', German imperialism was of a quite different order and represented the true 'world menace' (pp. 6–7). The Prussian government, another commentator declared, is the enemy of Christian civilisation, of moral progress, of spiritual enlightenment. In Prussia success becomes the standard of conscience, citizenship is prostituted to the lusts of militarism. 'A cannibal feeds on the blood of one man at a time. These militarists batten on the blood of thousands. For them manhood is not a divinely imparted life, but a demoniacally invented war-machine.'[22] Or as Kipling put it: 'We are dealing with animals who have scientifically and philosophically removed themselves inconceivably outside civilization.'[23]

German propagandists charged the British Empire with expansive and brutal territorial ambitions. Their British opponents replied that certain empires were, so to speak, natural and organic; such cases as the British Empire had to be distinguished from the aberrant history of Germany and above all from the 'peculiar origins of the Prussian state':

> [The British Empire] has slowly taken shape, during the last four centuries, since intercourse was opened up by sea between different races of the world as a whole. Its collapse, at the hands of Germany or any other Power, would not mean the substitution of a non-British Empire for a British. It would inaugurate a period of chaos in all five continents of the world.[24]

British naval power, according to these authors, is a fact not of politics but of natural history and 'evolution': 'Any Power that challenges the naval supremacy of Great Britain is quarrelling, not with the British government or the British people, but with the facts of history, of geography and of the political evolution of the world' (p. 112).

'The Prussian school' had erroneously understood the British Empire to be a matter of conquest where in fact it was founded not 'in a love of glory, but a sense of responsibility towards backward peoples' (p. 113).

When Allied propaganda in the First World War diagnosed a deep fissure in the German tradition, it deployed a racial vocabulary of nationality which, as I have shown, had already been elaborated in French and British anthropology after 1870. There were apparently two quite distinct racial genealogies in Germany. As Seton Watson reaffirmed in 1914:

22 Carliol, 'The inner meaning of the war', p. 736.
23 Kipling, *The War and A Fleet in Being*, p. 86.
24 Seton Watson *et al.*, *The War and Democracy*, p. 96.

6 'British Civilization' (from Patrick Ford, *The Criminal History of the British Empire*, 1915).

In order to understand Prussia and the Prussian spirit we must plunge ourselves into an atmosphere wholly different from that of the Germany that has just been described. The very names of the two countries mark the measure of the difference. Germany means the country of the Germans, as England means the country of the English. But the name Prussia commemorates the subjugation and extinction by German conquerors and crusaders from the west of the Prussians or Bo-Russians, a Slavonic tribe akin to the Russians.[25]

Prussia was at once the machine and the spanner in the European works, emerging as a deranged, aberrant freewheeling creature, a

25   *Ibid.*, p. 95.

cancer in the general peculiarity of German history which Western historians and social scientists would locate in the late nineteenth century, in the inter-war period and in the decades after 1945.

The question of who desired the war ran over into wider issues in wartime propaganda about the representation of national identity. The pamphlets which poured forth in and beyond 1914 had the function not only of vilifying the other, but of constructing a version of 'us'. An essential literary tradition was evoked, comprising, for instance, Shakespeare, Milton, Wordsworth and Tennyson.[26] Literature was to be a key 'weapon' at the imaginary 'Court of Civilization' to which the propagandist appealed; the literary heritage of each warring nation was viewed comparatively and competitively. Paul Vinogradoff for instance rebutted the German charge that the war constituted a struggle of 'civilization' against 'Muscovite barbarism', by pointing to the great figures of Russian art, science and history who might compete in any international battlefield of culture. 'A nation represented by Pushkin, Turgeneff, Tolstoy, Dostoyevsky in literature ... need not be ashamed to enter the lists in an international competition for the prizes of culture.'[27]

Literature was sometimes directly displaced into military and naval images. According to one particularly startling formulation early in the war:

> The Germans to-day have somehow got it into their heads that they are, before all other nations, a nation of poets. Can they compare with us? Let us put it into naval language. Their 'Grand Fleet' seems somewhat limited. Grant that they have one 'super Dreadnought', the 'Goethe', admittedly a fine and powerful ship; still she is hardly equal in guns or speed to the 'Shakespeare'.[28]

The propagandists appealed to the imagination as a national possession to be jealously guarded and differentiated from outsiders. The arts and philosophy, it seemed, could be mapped out in terms of national character. Each nation had its spiritual essence. Internal differences were screened out of the representation; the lines of conflict were treated as purely external. It was important to maintain that unity, never to lose those 'roots' or dilute that cultural essence and become 'Huns in our own country':

> The English character is rooted in history and tradition; it is rooted also in local associations – the beauty of English fields, the glories of sea and sky, and the charm of ancient buildings. Let us

26　See Matheson, *National Ideals* [P.], p. 27.
27　Vinogradoff, *Russia. The Psychology of a Nation* [P.], p. 9.
28　Warren, *Poetry and War* [P.], pp. 3–4.

remember that it lies with us to guard these beauties: not to be Huns in our own country – to defend it from outward dishonour and disfigurement. But deeper and stronger even than these incalculable influences which pass into our very substance through the associations of the senses, are those forces of imagination which for many of us control the deepest springs of our being.[29]

The independence and individualism of spirit which was often proposed as a singularly English virtue was rather a double edged attribute in wartime. One commentator, fearing perhaps that independence might quickly result in loss of solidarity, and apathy towards the war, insisted that in fact the unique English devotion to freedom and the individual reinforced the nation and its struggle:

[England's] people are not just now swayed by strong passions, moving her one way or the other. Centuries of freedom, won and maintained by the blood of our forefathers, have given us the habit of thinking and acting each for himself, and in doing so of bearing in mind the common good.[30]

English propagandists were faced with something of a problem: they sought to portray Germany as the more militaristic, bellicose, war-loving nation, thereby risking the assumption that this island people was dangerously pacific and even effete. As Wilkinson put it in *The Coming of the War*:

As Englishmen we are all at some disadvantage when we are called up to state our ideals. Our native inclination is to do the right thing and make no fuss. We are shy of using large language, and of claiming to be better than other people. We prefer to justify our action in off-hand language as 'playing the game' or to speak lightly of our responsibilities as 'part of the day's work'. (p. 1)

Allied propaganda frequently insisted that German aggression was innate. The Germans were simply born that way, or at least certain values had been inculcated from birth, as Kipling made clear: 'the Hun has been educated by the State from his birth to look upon assassination and robbery, embellished with every treachery and

---

29 Matheson, *National Ideals* [P.], p. 27.
30 Wilkinson, *The Coming of the War* [P.], p. 7. Or as another commentator carefully pointed out, the English were not defeatist, but strong and well controlled: 'Someone, I forget who, has said that we English are not a military nation, but that we *are* a very warlike and even pugnacious people. It is very true. There is no fear that we shall ever become militarist, but we *are* a fighting race' (Warren, *Poetry and War* [P.], p. 3).

abomination that the mind of man can laboriously think out, as a perfectly legitimate means to the national ends of his country'.[31]

The war, it was widely agreed, exceeded any utilitarian function. Its origins lay in the hidden pools of human instinct, an unconscious world of primitive drives, a will to power beyond the realm of politics, diplomacy and civilisation. The struggle was to assert the unique and superior national virtues which had emerged from the primeval pool of racial instincts. As one German propaganda message conceived it:

> The die is cast;...the race...called so appropriately 'the modern Greeks', is at last engaged in the long expected death-struggle with its jealous and semi-barbaric foes – a struggle that one [*sic*] begun, must go on, however often interrupted, to the bitter end, with passions as violent and instinctive as those of primitive man, and for issues that do not all lie upon the surface.... *For the death-grapple of Slav and Teuton is not merely a struggle for territory or for commercial supremacy, as so many superficial observers seem to believe, but a conflict of principles...[a struggle] of the highest ideals known to the human race against the low and sordid aims of races merely veneered with culture.*[32]

The war's causes ran deep, and its effects seemed to raise the participants towards the sublime. As A.L. Smith, Master of Balliol College Oxford, put it to an audience at King's College London in 1916, the war had created a sacred union and had put paid to England's damaging tradition of individualism:

> War is indeed a mighty creator. It is an intellectual awakener and a moral tonic. It stirs men to think, and thinking is what we most lack in England. It creates a conscious unity of feeling which is the atmosphere needed for a new start. It purges away old strifes and sectional aims, and raises us for a while into higher and purer air. It helps us to recapture some of the lofty and intense patriotism of the ancient world. It reveals to us what constitutes a modern nation, the partnership between the living, the dead, and yet unborn.[33]

Pacifist opposition to the war frequently sought to challenge the differentiation of 'us' and 'them' in terms of 'Prussianism'. War

---

31 Kipling, *Kipling's Message* [P.], p. 6.
32 Sanborn, 'Why the Teuton fights', p. 26.
33 Smith, 'The people and the duties of empire', pp. 43–4.

critics insisted that a condition of barbarous militarism existed on both sides. As the suffragist and pacifist Helena Swanwick put it:

> We British have invented the name of 'Prussianism' for a doctrine which we are finding very ugly and hateful. But we should not forget that it is the very doctrine with which our British Anti-Suffragists have made us very familiar during the past ten years and which has been enunciated even by the Prime Minister.[34]

All the great states, she observed, are 'Prussian' in so far as they are organised on a militarist basis. In a male-dominated world ('Men make wars, not women') conflict is generated by a combination of traditional conceptions of honour, the pursuit of material gain, the drive of vested interests, love of domination and what is often called glory, fear, indolence of mind, pugnacity, love of hazard and adventure, and disgust with the drabness of daily life (pp. 6–8).[35]

Norman Angell also noted the 'Prussianism' of the British Empire, that is to say, its triumph through strength, war, conquest, guile, mastery and so on.[36] He observed the curious double standard whereby Germany's anthem was deemed sinister, but not 'Rule Britannia'. Furthermore, Bernhardi's war-glorifying works in Germany were taken by British propagandists to typify the national spirit whilst, for instance, Spenser Wilkinson, the Professor of Military History at Oxford, could, with impunity, brazenly proclaim Britain's imperial mission of world domination.[37]

34 Swanwick, *Women and War* [P.], p. 5. Swanwick had been editor of the *Common Cause*, the journal of the National Union of Women's Suffrage Societies, between 1909 and 1912. She was a founding member of the Union of Democratic Control. Her critique of war was part of a wide-ranging feminist analysis of the link between militarism and the oppression of women which developed during the First World War. See for instance Florence *et al.*, *Militarism versus Feminism*.

35 Virginia Woolf extends that argument still more polemically, for some exorbitantly, in the late 1930s. In *Three Guineas* she argues that the patriarchal state *at large* operates with the same forms of sexual tyranny as European Fascism. In both, Woolf argues, the man aims to dictate the terms of the life of the woman, to determine the very definition of the woman: 'Are they not both the voices of Dictators, whether they speak English or German, and are we not all agreed that the dictator when we meet him abroad is a very dangerous as well as a very ugly animal? And he is here among us, raising his ugly head, spitting his poison, small still, curled up like a caterpillar on a leaf, but in the heart of England' (*Three Guineas*, p. 62; *cf.* p. 205 n. 48).

36 Sir Ralph Norman Angell Lane (1872–1967); author of *The Great Illusion* (1910) and numerous other writings on war and peace; recipient of the Nobel Peace Prize in 1933.

37 Angell, *The Prussian in our Midst* [P.], pp. 21–2.

Indeed as the war proceeded, observed Angell, the problem of distinguishing the antagonists became ever more difficult, propagandist rationales ever more threadbare, since war of itself produced, as it were, a 'Prussianisation' of everyday life in all parties:

> We may be fighting for democracy, freedom, parliamentary government, against despotism, government by a military caste, and restraint of free speech; yet, if we are to wage the war efficiently, our government must be autocratic, free speech must be suspended, and the military order must have arbitrary power. (p. 9)

The Allied press appealed to the 'war of civilization', but some journalists simultaneously spoke of the need to pursue the contest through to some ruthless absolute victory. Angell wryly noted how a recent newspaper article, discussing the prospects of an Allied entry into Germany in September 1914, advised against moderation and urged the army to pursue the 'Hun' mercilessly:

> The one thing that Germany needs is the infliction of a pitiless punishment – one that will bring home to every citizen in the German Empire his personal share in all the devilries that have been committed, and make him realise that the task of civilization is being applied to his back by way of requital.[38]

The war was rationalised as the inevitable unleashing of instincts and passions, which all too easily 'releases us ... from the categoric imperative, which in our relations with individuals imposes upon us the obligation to check impulses, to control our temper' (p. 19).

Many of the pacifist writers in this period abhorred the racial posturing, the jingoism and militarism, the crude appeals to bellicosity as 'human nature'. Nevertheless, we sometimes find them redeploying the same racial anthropological and evolutionist language, as though the words 'primitive', 'savage', 'barbarous', 'civilized', 'cultured' were quite neutral, beyond analysis or reproach. Bertrand Russell believed not only that he was witnessing 'a return to the savage beneath the miserable rags of a tawdry morality',[39] but also that he was witnessing a racially determined Eastern atavism. The Western war apparently had not the same 'ethnic inevitability' as the Eastern. Germany's conflict was in substance defensive, 'the attempt to preserve Central Europe for a

---

38 Quoted in *ibid.*, p. 26.
39 Russell, *War: The Offspring of Fear* [P.], p. 3.

type of civilization indubitably higher and of more value to mankind than that of any Slav State' (p. 4).[40]

> Essentially this war...is a great race conflict, a conflict of
> Teuton and Slav....The conflict of Germany and Russia has
> been produced not by this or that diplomatic incident, but by
> primitive passions expressing themselves in the temper of the two
> races.

> The Austrians are a highly civilized race, half-surrounded by
> Slavs in a relatively backward state of culture....Servia, a
> country so barbaric that a man can secure the throne by instigat-
> ing the assassination of his predecessor, is engaged constantly in
> fermenting the racial discontent of men of the same race who are
> Austrian subjects.[41]

Compare these remarks with the view of a pro-German
propaganda paper distributed in the United States to set out the
'Teutonic' case:

> Anybody familiar with the perfidy and the atrocities of the
> Balkan States in their two recent wars or with their bestial
> murder of their own king and queen can certainly realise that we
> have to do here with races living on the plane of semi-savagery –
> a people impervious to reason.[42]

That perception of the inferiority and separateness of the Slavs was
frequently shared by both English and German propagandists. As
one English commentator said:

> To back our Western friends in a war of defence is one thing, to
> fling ourselves into the further struggle for the Empire of the East
> quite another. *No call of the blood,* no imperial calculation of

40 On 1 August 1914 *The Times* published a letter, headed 'scholars' protest
    against war with Germany', with nine signatories, mostly from Cambridge. It
    declared: 'We regard Germany as a nation leading the way in Arts and
    Sciences, and we have all learnt and are learning from German scholars. War
    upon her in the interests of Servia and Russia will be a sin against civilization'
    (*The Times,* 1 August 1914, p. 6). The same page also carried a letter by
    Norman Angell which viewed war with Germany as folly, because it would
    involve allying Britain with Slavic backwardness. Was it not a frightening
    mistake, Angell asked, to identify with Russia and her Slavonic federation of
    say 200 million autocratically governed people, 'with a very rudimentary
    civilization', rather than 'a dominant Germany of 65,000,000 highly civilized
    [people] and mainly given to the arts of trade and commerce?' Both these
    letters of course were rebutted by various other correspondents.
41 Russell, *War: the Offspring of Fear* [P.], pp. 3–4.
42 Sanborn, 'Why the Teuton fights', p. 27.

self-interest, no hope for the future of mankind requires us to side with the Slav against the Teuton.[43]

Or as the title to a German propaganda piece put it 'fraternally' to its American readership, 'The present and future civilization of the world politically lies in the hands of the three great Teutonic states, Germany, England and the United States'.[44]

It would be difficult to overestimate the centrality of the notion of 'civilisation' in the language of the First World War. A broad distinction between 'civilisation' and 'barbarism' was used to distinguish the European imperial powers from their colonies; at other times to differentiate sections of the domestic population within a specific state; alternatively 'civilisation' was deployed to contrast the behaviour and the genealogy of one European nation with another. But the differentiation of the civilised and the uncivilised so often turned specifically upon nineteenth-century conceptions of evolutionary advance and backwardness. In the following passage from a speech in 1911 by the British Foreign Secretary Sir Edward Grey, for example, military and naval expenditure are viewed as an inevitable function of the discrepancy between the 'civilised' and the 'less advanced'. But in a 'paradox' which needs to be drummed home, the arms race threatens to undermine and destroy the 'civilisation' it defends, since the pursuit of new weaponry does not simply mark some collective 'civilised' superiority over the 'uncivilised', but exacerbates rivalries within Europe:

Some naval and military expenditure I admit the most highly *civilised* nations must necessarily have until the world is all equally *civilised*. The most *civilised* nations must, of course, have in all circumstances the power to protect themselves against those who are *less advanced*. But the paradox remains, that their expenditure on armaments is not directed against nations *less civilised* than themselves, not against *more backward* nations, but is directed – I will not say directed against, but it is entered upon in rivalry with each other. The paradox – unless the incongruity and mischief is brought home not only to men's heads generally, but to their feelings, so that they resent the inconsistency and realise the danger of it – if this tremendous expenditure on and rivalry of armaments goes on it must in the long run break *civilisation* down.[45]

43 Brailsford, *The Origins of the Great War* [P.], pp. 16–17; emphasis added.
44 Burgess, 'Present and future civilization of the world', p. 5.
45 Grey, *The Peace of the World* [P.], p. 9; emphasis added.

The felt impossibility of casting Germany as simply culturally underdeveloped or backward led to more subtle and complex pejorative distinctions in First World War rhetoric. Germany, it seemed, suffered a socio-biological perversion, a racial crack, a degeneracy stemming from over-rapid development. The Germany of high culture was counterposed to another Germany, now laid bare as the nightmare 'monstrosity' within Western civilisation, a parasitic state, a vampire nation. There was said to be a radical and inalienable difference inside European civilisation and a split in the European psyche. As Ramsay Muir summed up in 1915:

> perhaps the worst nightmare of this war has been that it has produced, or rather has revealed to us, the existence of an extraordinary gulf yawning between the nations, between Germany on the one hand, and the Allies on the other. The standards of judgement, the ideas of right and wrong, entertained by the two sides in this war, are so wide asunder that each is totally incapable of understanding the point of view of the other.[46]

It was difficult to explain this gulf. The Germans after all were not 'down-trodden or semi-barbarous'. Muir admitted to 'helpless bewilderment' and asked: 'Can a whole nation be fundamentally immoral? That would be a contradiction of the very idea of morality. Can a whole nation suddenly go mad . . . ?' (p. 258). The putative madness of Germany was not a function of its backwardness, but of its strange technological success and excess. This brings us back to a central issue in the present study – the complex interlinking of monstrosity, machinery and modernity. German efficiency and science, Muir suggests, has fostered a strange failure of sympathy and understanding, a profound incapacity to grasp the feelings of 'non-German minds' (p. 263). Muir is at pains to downplay the 'racial affinity' between the German and the English. The anthropological constitutions of the two nations, we learn, are decidedly different, as any would-be English customer in a German hat shop would rapidly discover:

> In our very composite people the Teutonic is but one element among many, though it has given us the framework of our language; while the Germans themselves have a large inmixture of Slavonic blood, in which we have no share. An Englishman need only try to fit his head with a hard hat in a German hatshop to realise how different is the skull-formation of the two peoples. (p. 264)

46 Muir, *The Antipathy between Germany and England* [P.], p. 258.

Whilst atrocity stories played a major part in the propaganda released by both sides during the First World War, the Allies particularly linked the enemy's outrages to the machine-like mentality which the Germans had acquired in the course of the nineteenth century. Atrocity had been elevated to a science and a technology. German commentators on the other hand frequently reviled the Allies' barbarity, speaking of how the eyes of 'wounded German soldiers in Belgium were gouged out, and their ears and noses cut off'.[47] Allied propagandists spoke consistently of a brutal and systematic Teuton ruthlessness; 'science' itself had been perverted and degraded.[48] Redressing the monstrous atrocities committed by the other side became in its turn a rationale for the war. Real atrocities, in defiance of all protocols and conventions, undoubtedly were committed within that wider atrocity, the war itself, but the imagery also began to generate its own momentum, its own spiral of new fantasy and horror which coalesced with pre-war emblems of torture and cruelty.

The notion of the First World War as absolute historical schism was a powerful structuring assumption of wartime propaganda. There is a complex relationship between the official self-presentations of the European nation states during the war, wider and earlier cultural commentaries, and subsequent historiography. The year 1914 appeared to many contemporary commentators to constitute a definitive rupture from, or in some formulations, a fratricidal civil war within, a common European civilisation.[49] War

47 *The Fatherland, Truth about Germany*, p. 21.
48 As the National War Aims Committee put it in 1918, 'In the verified record of the deeds done then in Belgium we see savagery elevated to a science in pursuance of the war aim of domination. The cold catalogue is enough: a baby crucified with hands and feet outstretched, nailed like a rat to a barn; another baby carried aloft, skewered on a bayonet in a regiment of singing soldiers; girls violated again and again until they died; matrons, old men and priests slaughtered; men mutilated in ways that one man can hardly whisper to another; women and children thrust forward as a screen between "the gallant troops of Germany" and their enemy; organised massacre; the abuse of the Red Cross and the White Flag. Everything that we thought secure among civilized man was defiled and destroyed – fidelity to the pledged word, reverence for age, the sanctity of womanhood, childhood and weakness; standards of honour, of justice and of clean fighting. And they were destroyed, not in an access of passion, but on a deliberate and calculated policy of "frightfulness"' (NWAC, *Aims and Efforts of the War* [P.], p. 47).
49 As the American racist theorist Lothrop Stoddard was to comment in 1920: 'To me the Great War was from the first the White Civil War' (*The Rising Tide of Color against White World-Supremacy*, p. vi). Stoddard's book is haunted by the swamp of blackness, the tides of the non-white world overcoming the Caucasian 'dikes'. The First World War is 'the white death

was perceived as a breaking apart of a hitherto shared tradition; or conversely as the exposure of that tradition as lie or illusion. For some, the First World War was the origin of a new vicious circle whose end was unforeseeable.

It became a convention to speak of the First World War as 'a war of ideas – a conflict between two different and irreconcilable conceptions of government, society, and progress'.[50] It was apparently a war over the very conception of civilisation and culture (or *Kultur*): 'Kultur is a difficult word to interpret. It means "culture" and a great deal more besides'; 'Germany thinks of civilisation in terms of intellect while we think of it in terms of character' (p. 352–3). Allied propaganda argued that there was a radical fissure in the German tradition: a deep and immutable split lay between the thought realms of Kant and Hegel, and each philosopher was taken to represent a side of German national character: 'If Kant is the philosopher of one side of Prussia ... Hegel is the philosopher of another side'.[51]

Not all Germans desired the war, one British pamphlet argued in 1914; military bellicosity was 'really at bottom Prussian rather than German'.[52] The course of nineteenth-century German history and the specificity of Prussia's role in that history slid over into

grapple ... [which] merely accelerated a movement already existent long before 1914' (p. 15). 'The heart of the white world was divided against itself, and on the fateful 1st of August, 1914, the white race, forgetting ties of blood and culture, heedless of the growing pressure of the colored world without, locked in a battle to the death. An ominous cycle opened whose end no man can foresee. Armageddon engendered Versailles; ... The white world to-day lies debilitated and uncured; the colored world views conditions which are a standing incitement to rash dreams and violent action' (p. 16).

50   Seton Watson *et al.*, *The War and Democracy*, p. 348.

51   Barker, *Nietzsche and Treitschke* [P.], p. 4. *Cf.* Edward Timms's remark: 'To foreign observers (as to Kraus himself) there seemed to be "two Germanies": the Germany whose art and music, philosophy and literature had captured the admiration of the civilised world; and the Germany whose armies were apparently hell-bent upon its destruction. The distinction is made above all by western writers well-disposed towards Germany. Both Romain Rolland and Bernard Shaw draw a sharp line between Prussian imperialism (which must be destroyed) and the German people with their humanistic traditions (which must be protected and preserved). This explanatory model of Germany may seem simplistic. It ignores that "third Germany" which constitutes the missing link: the advanced industrial society whose economic prosperity both financed the programme of military expansion and funded the intellectual. But the model was advanced as a more enlightened alternative to the crude anti-Hun propaganda which followed the German invasion of Belgium' (Timms, *Karl Kraus*, p. 305).

52   Sanday, *The Deeper Causes of the War* [P.], p. 6.

an anthropology of German national character which pinpointed national atavism, degeneracy, splitting and perversity.

In his popularisation of the Bryce Report on German atrocities in 1916, J.H. Morgan cited the French writer Fustel de Coulanges (1830–89) on the unusually high incidence of homosexuality in Berlin. The Germans apparently showed atavistic characteristics typical of the original Huns. As Morgan went on, there might even 'be force in the contention of those who believe that the Prussian is not a member of the Teutonic family at all, but a "throw-back" to some Tartar stock'. Germany was a 'hybrid nation' with the acquired 'idiom' of Europe and the 'instincts of some pre-Asiatic horde'. As Morgan wrote in his report on *German Atrocities* in 1916, the enemy nurtured 'dark atavisms and murderous impulses' beneath a civilised veneer'.[53]

German propagandists, it should also be noted, sometimes argued that there were two Englands – Anglo-Saxon (benign) and Norman (pernicious, devious, cunning): 'The English nation is composed of two strong racial elements – the Anglo-Saxon, honest, trustful, outspoken and liberty-loving; and the Norman-French, taciturn, enterprising in the cause of conquest, ruthless in the employment of brute force and expert in hypocritical subtleties.'[54] The enemy was sometimes cast as fearsome because monolithic, and at other times because plural and polymorphous. The Germans accused the English of a kind of treacherous miscegenation, forging an alliance with black- and yellow-skinned people, polluting the very purity of England, blurring the lines of division between the European and the non-European, or between the superior and the inferior Europeans. German commentators played on the shared Anglo-Saxon roots of the two nations and set this union against the Slavs – hence they derided England's unholy alliance with a 'backward', 'degenerate' and monstrously fecund inferior people.[55] 'The great issue,' it was insisted, 'has been and is now whether the Slav is to rule from the Japanese Sea to Berlin and further West, or whether Germany, even fighting with her civilized Western neighbours, is to stand up to maintain European civilization and save it from the Rule of the Knout.'[56] It was confidently predicted

53 Quoted in Wallace, *War and the Image of Germany*, p. 183. Cf. 'the soul of Germany was being eaten away by the virulent poison of Prussianism' (Kahn, *The Poison Growth of Prussianism* [P.], p. 18).

54 Schrader, introduction to *England on the Witness Stand* [P.], p. 3.

55 '[The Slavs'] rate of increase is nearly three times that of even the Germans' (*The Fatherland, Germany's Just Cause*, p. 20.).

56 Dernberg, 'Germany and the war', p. 18.

that 'A revulsion in favour of the Saxon may almost certainly be counted on before the war ends'.[57]

In Allied propaganda, then, Germany was split into two parts, but these were not viewed as historically constant. Nineteenth-century history was the saga of the *accumulating* power of the darker tradition. As the journalist and social theorist L.T. Hobhouse observed in 1916: 'Follow German thought down through the century, and you will only find more and more insistence on force, power, ascendancy, and more and more repudiation of any binding law or any general sense of the unity of mankind.'[58]

Hobhouse rejected the pacifist argument that both sides had been reduced to the same level. The crisis was external – 'a calamity that has befallen us from without, not the corruption from within of which nations perish' (p. 31). Nevertheless Hobhouse asked himself to what extent the world's future was now caught up in an indefinite dialectic of war and armed truce. Against that vivid and terrifying prospect, he could only hope for a new internationalism, and the achievement of the ultimate ideal of a united Europe (p. 188). But the hope was to prove stillborn.

War, Hegel had declared, is necessary to the constitution of the nation. To put it in another way, violent recognition of the imagined 'non-community' is essential to collective identity. I allude here to Benedict Anderson's discussion of the nation as an 'imagined community'. Whatever the complexity of relationships and the depth of exploitation within a particular nation, it is always 'imagined' as 'a deep, horizontal comradeship'.[59] Equally, of course, it is always constituted in relation to other states beyond its own frontiers.

Nationalism involves a kind of faith, indeed something of the aura of 'religion' which it arguably both displaces and substitutes. The war dead are central to nationalism. Anderson points to the striking modern emblem of the cenotaphs and tombs of unknown soldiers which, he argues, have no true precedents in earlier times. These tombs, 'saturated with ghostly *national* imaginings', are tied to nationalism, and are part of the surrogate religious bridge which, unlike either Marxism or liberalism, nationalism succeeds in providing between death and immortality. Nationalism begins to emerge in eighteenth-century Europe; precisely in that epoch of crisis and decline for orthodox religious belief, it constitutes a secular trans-

57   *The Fatherland, Germany's Just Cause*, p. 10.
58   Hobhouse, *Questions of War and Peace*, p. 20.
59   Anderson, *Imagined Communities*, p. 6.

formation of fatality into continuity, contingency into meaning, the ephemeral into the eternal.

Across the nineteenth century the meaning of nationalism shifts. In his recent study, *Nations and Nationalism since 1780*, Eric Hobsbawm traces the 'transformation' of nationalist ideology and image. Broadly, it is argued, nationalist ideology moves from an expansive ethos towards a more constricted, excluding and often racially determinist conception of collective identity. This process gathers momentum from the mid-century, but comes of age after 1870. Conversely, others have argued that this double aspect, traced chronologically by Hobsbawm, was already powerfully inscribed within the French Revolution itself, and perhaps even earlier both aspects had been inextricably intertwined within the contradictions of Rousseau's writings.[60]

Whether one is speaking of the French Revolution in particular, or more abstractly of the nature of 'imagined communities', the discussion needs to turn precisely on the imagined non-community, both inside and outside of 'this' body, 'this' corporate entity. The Revolution still constitutes the touchstone of an ethical universalism, an expansive internationalism, not least with its adoption of various foreign writers as French citizens. But whilst the Revolution appealed across borders to 'humanity', it also contained increasingly powerfully a particularist drive, a gathering rejection of the foreigner.[61]

During the 1790s, the army was to be a key site of this drive to purge France of aliens. In April 1793 Robespierre demanded

60 The argument is made, for instance, in Kristeva's *Etrangers à nous-mêmes*, p. 257. *Cf.* Guiraudon, 'Cosmopolitanism and national priority: attitudes towards foreigners in France between 1789 and 1794'.

61 A decree was adopted on the 26 August 1792 conferring the title of French citizen on various foreign savants who had furthered reason and liberty – amongst them Priestley, Paine, Wilberforce, Bentham, Washington and Schiller. The initial hope was that the principle of the Rights of Man would be taken home from France by the foreigners who had been welcomed by the Revolution; the wish that such ideas would then penetrate the consciousness of the masses causing insurrection against absolutist tyrants gradually gave way in the face of war to the view that foreigners were infiltrating something highly damaging *into* France; and even that the enemy had directly sent in spies and saboteurs. On 18 March 1793 Barère produced on behalf of the Comité de Salut Public a repressive law against strangers. Cambon demanded that all outsiders be made to leave the territory of the Republic: see Kristeva, *Etrangers à nous-mêmes*, p. 233. 'Cosmopolitan' ideas and defenders, like the Hébertistes, were increasingly beleaguered. Foreigners, deemed responsible for the intensifying economic problems of the Republic, faced new difficulties. They were subject to being stopped, sometimes arrested, interned, even executed.

the expulsion of foreign generals 'to whom we have imprudently entrusted the command of the army'.[62] In the long wake of the Revolution both in France and throughout Europe, armies were increasingly 'nationalised' – their personnel were now to be drawn from the country for which they were fighting.[63] During the Napoleonic wars, 'nationals' continued to fight on both sides although the liaison between birth and military affiliation was to become increasingly common.[64]

In *Citizens* (1989), Simon Schama explores in some detail this intersection of militarism, xenophobia and 'fifth column conspiracy panic' within the Revolution. Whilst there were real external threats, Schama insists that patriotic war and the spiral of violence were not by-products, accidents or simply external phenomena, but intrinsic to the very logic of 1789 and its 'philosophical universalism'.[65] 'Militarised nationalism,' he insists, 'was not, in some accidental way, the unintended consequence of the French Revolution; it was its heart and soul' (p. 858). Schama charts the shifts between 'amiable cosmopolitanism' and 'crusading self-righteousness', the consolidation of a Romantic discourse of blood, honour and soil as the arousing hymn of Revolution.[66] His descriptions of the language and the practice of 'weeding out' traitors, above all in the journalism and the politics of Marat, or the 'domiciliary visits'

62 Quoted in Kristeva, *Etrangers à nous-mêmes*, p. 234. By 25 December 1793 Robespierre was accusing foreigners of causing all France's crises; see *ibid.*, p. 236.

63 As Anderson remarks: 'While the armies of Frederick the Great (r. 1740–86) were heavily staffed by "foreigners", those of his great-nephew Friedrich Wilhelm III (r. 1797–1840) were, as a result of Scharnhorst's, Gneisenau's and Clausewitz's spectacular reforms, exclusively "national-Prussian"' (*Imagined Communities*, p. 28). Cf. Hobsbawn, *Nations and Nationalism since 1780*.

   Whilst the movement towards the national army was largely a phenomenon ushered in by the French Revolution, Machiavelli had already moved in this direction with his critique of the disloyal, rapacious, treacherous, impudent and cowardly *condottieri* on whom he perceived Florence to be so cripplingly dependent (see Pocock, *The Machiavellian Moment*; Gat, *Origins of Modern Military Thought*).

64 Heinrich Brandt (1789–1868) provides one example – a Prussian soldier who served under Napoleon (in the Grand Army of the Dukedom of Warsaw, 1807–13), he rejoined the Prussian army in 1813. See Rapoport's introduction to Clausewitz, *On War*, p. 19. On the 'cosmopolitanism' of eighteenth-century British army and navy officers, see Brewer, *The Sinews of Power*, p. 55.

65 Schama, *Citizens*, p. 592. Cf. 'patriotic war was, in fact, the logical culmination of almost everything the Revolution represented' (*ibid.*, p. 591).

66 *Ibid.*, p. 593.

organised by Danton, where it was proclaimed that 'Everything belongs to the *patrie* when the *patrie* is in danger',[67] are deeply evocative. Beyond the immediate exigencies of the 1790s (which were real enough), Schama explores the more complex ideological links between the Terror, militarism and the modern cultural-political quest for purity and a radical unmasking of the false within the true. He argues that what gave the *sans-culottes*' demands special force in 1792 was the dynamic of military patriotism. The enemies within were now not some abstractly defined class foe, but, so to speak, 'Austrians in French dress'. It was said at the time that the menacing 'Austrian Committee' causing so much chaos and demoralisation on the front was also actively and coldly provoking moral and economic crisis at home, and most tangibly of all, that it lay behind the food shortages. There was, as Schama tells us in his 'chronicle', a continuing 'craving' to identify, unmask and punish the fifth-column patriot-hypocrites in the Jacobins and the Cordeliers.

It is not my aim to review Schama's vivid, compelling and deeply engaged history of the Revolution's developmental pathology: 'From the very beginning – from the summer of 1789 – violence was the motor of the Revolution.'[68] Rather I want to draw attention to that 'violent motor' image which has been so central a figure in the nineteenth-century material discussed in the preceding chapters. The notion of a vast and ghastly sanguinary spiral across the nineteenth century, stretching perhaps from 1789 to 1914, or even to Auschwitz (as Schama occasionally obliquely suggests), a 'dialectic of the enlightenment' (to borrow the Frankfurt school title), of blood and madness, of reason and violence, has been theorised in many ways in the past. From Burke to Carlyle to Taine and beyond, the French Revolution was understood as both apotheosis and involution of the Age of Reason.

Even within traditions sympathetic to Marxism, the story of revolution, class war and imperialism has provoked for a long time a profound hesitation, an insistence on a dynamic of war and the irrational which cannot be explained away in a language of interests. Take the critique of Joseph Schumpeter, German economist, author of the influential study *Business Cycles* (1939) and Professor of Economics at Harvard from 1932. In Schumpeter's study of imperialism, Schama's 'motor' of revolutionary violence was theorised as the legacy of the ferocious persistence of an economic and psychic *ancien régime*, a bloody realm of aristo-

67 Quoted in *ibid.*, p. 626.
68 *Ibid.*, p. 859.

cratic desire which would not let go of the fundamentally pacific bourgeoisie.

Lenin's view that imperialism and its ensuing wars represented a kind of terminal point of capitalism was thus to be disputed in Schumpeter's important critique. According to his argument, it is a simplification to see imperialism as a direct outgrowth of the conditions of capitalism. He reaffirms Marx's view and indeed Cobden's, that capitalism is the very antithesis of the warrior creed. Capitalism, he insists (echoing the *Communist Manifesto*, that 'glowing tribute' by Marx, as Schumpeter puts it, to bourgeois economic relations)[69] is fundamentally anti-heroic: 'the ideology that glorifies the idea of fighting for fighting's sake and of victory for victory's sake understandably withers in the office among all the columns of figures' (p. 128). Indeed Schumpeter insists that 'the industrial and commercial bourgeoisie is fundamentally pacifist' (*ibid.*). This is not to say that historically a bourgeois system or grouping will not condone war where it pays, but the logic of bourgeois capitalism, the condition of its productivity and of its mentality, favours peace. It is in this sense that the 'Marxist theory that imperialism is the last stage of capitalist evolution therefore fails quite irrespective of purely economic objections' (p. 129). The history of capitalist stages is understood here in terms of the shift from an early 'romantic' phase (a time of genius, speculative courage and individual invention) to an era of the expert, the official, the calculable, the routine – an age of mechanical reproduction in which 'trained specialists' turn out predictable produce, statistics and results. Peace is simply the best medium, the relatively frictionless milieu, in which capitalism develops.

In the field of war, Schumpeter suggests, the watershed between the 'heroic' and the 'specialist' age lies between the eighteenth and nineteenth centuries. Whilst from the Middle Ages onwards, he believes, warfare becomes ever more mechanised, up to and including the Napoleonic wars it continues to turn decisively on individual decision and the driving power of the leading man. Napoleon's existence and his presence at battles were crucial. Subsequent rationalisation and specialisation, however, have continued to 'blot out personality' (p. 133).

Imperialism and its attendant adventurism should not be seen as the leading principle of the modern age, we are told. On the contrary they are fundamentally 'atavistic' in character. But whilst they may be primitive, they are profoundly tenacious. What Schumpeter seeks to locate is the persistence of something 'irrational', inimical to

69   Schumpeter, *Capitalism, Socialism and Democracy*, p. 7.

economic 'interests', a powerful dynamic presence inadequately theorised by Marxism; aggression and expansion not as means to an end but as ends in themselves: 'Such expansion is in a sense its own "object," and the truth is that it has no adequate object beyond itself. Let us therefore, in the absence of a better term, call it "objectless".'[70] Far from being the realisation of capitalism, this insatiable, 'objectless' hunger for expansion represents a 'survival' of something historically and economically ancient.

Whilst Britain is supposedly less encumbered than some other European nations by this aggrandising lust, Schumpeter nevertheless dates a new and irrational imperialist policy back to Disraeli and the early 1870s. This policy, with all its associated symbolism, is not merely a reflection of economic interest and profit motives. Nationalism and imperialism have a quite different psychopathological register irreducible to the world of 'reason':

> All other appeals are rooted in interests that must be grasped by reason. This one alone [national sentiment] arouses the dark powers of the subconscious, calls into play instincts that carry over from the life habits of the dim past. Driven out everywhere else, the irrational seeks refuge in nationalism – the irrational which consists of belligerence, the need to hate, a goodly quota of inchoate idealism, the most naive (and hence also the most unrestrained) idealism. (p. 14)

Tracing this drive back through the ages (to Egypt, Assyria, Persia) Schumpeter locates the ancient emergence of a distinguishable warrior caste and of a 'new social and political organization [which was] essentially a war machine' (p. 33). This was extremely difficult to dislodge and eradicate. Such an organisation had an insatiable appetite for war:

> Only in war could it find an outlet and maintain its domestic position. Without continual passages at arms it would necessarily have collapsed. Its external orientation was war, and war alone. . . . *Created by wars that required it, the machine now created the wars it required. (Ibid.)*

What this work so vividly evokes is a set of problems about the relationship between modernity and military conflict to which we repeatedly return in the present study; between some putatively ancient war drive and its pacific overlays; between 'interest' and wanton aggression; between political control and a violent machine that moves of its own accord.

70 Schumpeter, *Imperialism and Social Classes*, p. 7.

There are at least two stories being told here. In the one, as capitalism continues to triumph, wars of conquest or adventurism will recede and institutions like the Hague Court of Arbitration will flourish: 'A purely capitalist world therefore offers no fertile soil to imperialist impulses' (p. 90). Industrial workers, peasants and socialists are always anti-war, Schumpeter insists. The future is on the side of peace. In the other story the recalcitrant 'nobility' succeed in warping the bourgeoisie, distorting 'the inner logic of capitalism itself (p. 128). The past colonises the present and the future, imperialising them, turning them from cosmopolitanism to national sentiment, from business to the war machine. At stake here is not simply 'the persistence of the old regime', but a perverse strangle-hold of the irrational upon the bourgeois mind, a fatal liaison between supposedly separate economic and symbolic orders, in which 'old' sovereignty successfully co-opts the new: 'essentially imperialist absolutism has patterned not only the economy of the bourgeoisie but also its mind – in the interests of autocracy and against those of the bourgeoisie itself' (p. 125).

This more disturbing tale is not altogether convincingly reined back to the reassuring narrative of the first story, although not for want of trying; Schumpeter so graphically stresses the recalcitrance of this awful war-crazed hybrid that his insistence on its transient nature sounds hollow, an appeal rather than a clinching argument that in the long run history must witness the decline of militarism.

12 The Rationalisation of Slaughter

TIME AND MOTION

I have argued that the idea of war as a machine, which develops so powerfully in the nineteenth century and which crystallises above all in relation to Prussia in the 1860s and German unification in 1871, has two decisively significant and contradictory permutations in modern cultural representation. In the first scenario, war is portrayed as increasingly under the sway of the logic of technology, science and planning. In such a view, abstract strategy is modulated by empiricism, a practical recognition of the 'friction' of war. The course of the conflict does not follow the conscious will of the commander. War does not orbit around the army chief's mind any more than the sun revolves around the earth. This insistence endures in the present century and is often anchored in the 'moral' of the Somme. It is there that 'friction' reaches its sad apotheosis. But if war exceeds the conscious intentions of its own authors in practice, nevertheless in this first scenario the possibility of conceptual mastery of the riddle of war remains the prize and the ambition. War may not be controllable by one man on the battlefield, but it can still, in principle, be thought and written about definitively. War would be a natural science, albeit of great complexity.

'In a small way I am trying to do for war what Copernicus did for astronomy, Newton for physics, and Darwin for natural history,' J.F.C. Fuller declared in *The Foundations of the Science of War* (1926). 'My book is the first,' he claimed grandiosely and absurdly, 'in which a writer has attempted to apply the method of science to the study of war.'[1] Fuller insists on the historical indispensability of war.[2] Science (or rather some cross-references between cell theory, racial anthropology, assorted statistics, technological analyses and Gustave Le Bon's crowd studies) is deployed

---

1 *Foundations*, p. 18.
2 'Great nations are born in war, and decay in peace. All things strong, virile, and manly spring up during a great war; and only a few years back we saw among ourselves a whole empire gathered together to meet a common foe, each soldier possessed by one common thought – the conquest of the enemy even at the cost of his own life' (p. 130).

to move beyond a mere narrative of battles, to tap the supposedly deep sources of the war machine: 'We require not merely a chronology of past events, but means of analysing their tendencies – means of dissecting the corpse of war, so that we may understand its mysterious machinery' (p. 21). It is always 'the primitive in man', 'born of the first mother', to which Fuller's science of war and aggression appeals for its ultimate explanation:

> Behind the soldier there stands this mystical impulse, born of the first mother, born of the first protoplasm which, dividing, lost its individuality, its desire to live, so that its species may survive.
>
> It is this impulse which impels the soldier to do certain things so that his race may continue and prosper. (p. 131)

In the second scenario, however, the reassuring (Clausewitzean) image of human control, science, politics, instrumentality, racial improvement through war, is overthrown in a more nightmarish image of crisis. War is no longer the stable object of representation, but the threat of a more drastic foreclosure of meaning. War is internal and external; the catastrophic destroyer of its own author. As even George W. Crile's 'mechanistic' account of war and peace concluded in 1916:

> Whatever the future may bring, however, man to-day betrays at every turn that he is in reality a red-handed glutton whose phylogenetic action patterns are facilitated for the killing of his own and of other species; that with all of his beneficent control of the forces of nature, he has created also vast forces for his own destruction, so vast that civilized man is to-day in a death struggle with a Frankenstein of his own creation; that, although he controls a world of limitless force, and endless machinery, he yet fails to control that all-important mechanism – himself.[3]

From Clausewitz in the 1830s to J.F.C. Fuller in the 1930s, the enduring problematic is how to retain rational human control over the practice of war, how to prevent the collapse of a scientific discourse on war. But what also emerged in the nineteenth century, above all in Clausewitz himself, was an insistence on the need to rethink the reality of war on the basis of contemporary history and above all the Napoleonic wars. 'The farther we go back,' he claimed, 'the less useful becomes military history, as it gets so much

---

3 *A Mechanistic View of War and Peace*, pp. 95–6. Crile sees man as a psycho-chemical mechanism which becomes 'seasoned' to war (if not driven insane). War involved a mobilisation not just of the nation but of each individual body, although the resulting state was liable to be unstable (see p. 11).

more meagre and barren of detail. The most useless of all is that of the old world.'[4]

It is true that even in the nineteenth century there still exists a tradition of war theory characterised, in Michael Howard's phrase, by 'extreme formalism'. Mathematical theories of strategy and campaigns were carried by some 'to lengths of rococo absurdity'.[5] Certainly the notion that enlightenment rationalism in general and the model of the natural sciences in particular were fundamental to the understanding of war endured from eighteenth- to much nineteenth-century theory. Clausewitz remained a key point of reference, the seminal figure in the establishment of a positivist lexicon and dream of war where even 'friction' seemed reducible to laws.[6]

War, it was hoped, might be placed safely under the aegis of science. Again the language and the material circumstances of such an aspiration interact in complex ways – we are not talking of some hermetically sealed-off realm of 'ideas' with no connection with actual 'hardware' developments in the period. There were new conditions of science, engineering and technology; not simply new methods of reading an unchanging material world. But the new developments in, for instance, measurement, synchronisation, powers of replication and mechanical reproduction, which reached a kind of apogee around the *fin-de-siècle*, were never untouched by the contemporary discourses which had both partly prompted and subsequently made sense of them.

The railway transformed the experience of space, opening up new systems of communication, cutting across (or sometimes bearing no apparent relation to) the traditional 'road map'. Contemporaries speculated on the implications of the railroad for war and peace, for society and industry. The Prussian army used the railway to transport its troops, with decisive results both against Austria and against France. In addition, the telegraph had facilitated coordination across its separated divisions.[7] Contemporaries divided as

---

4 Clausewitz, *On War*, p. 237.
5 Howard, *Studies in War and Peace*, p. 25.
6 See Irvine, 'The French discovery of Clausewitz and Napoleon'.
7 On the challenge which all this posed to Antoine Henri Jomini's 'enlightenment inspired' and influential war theory (now criticised for its mechanistic and geometric formalism), and more generally on the controversy which broke out in Germany over whether the Moltkean strategy, the new military model, introduced a new order of war, fundamentally different from the Napoleonic system and now based on the superiority of exterior lines, see Gat, *Origins of Modern Military Thought*, p. 121. On the general decline of the influence of Jomini (1779–1869), quickly in Germany and France, more slowly in the United States and Britain, and the revaluation of Clausewitz around 1870, see *ibid.*, pp. 129–30.

to whether the railroad heralded a new human intercourse that rendered war impossible, or created new opportunities for military speed and power.[8]

But within that wider industrial and modernising transformation epitomised by the advent of the railway, we can also locate a further massive 'rationalisation' around the turn of the century. Space and time were being materially and conceptually transformed. Take the introduction of globally coordinated time as one case of nineteenth-century rationalisation – a process which was to have powerful consequences for industry, communications and war. Time was standardised in response to economic and military exigencies. As one Canadian engineer and pioneer in the promotion of uniform time imperialistically announced in 1886, the use of the telegraph 'subjects the whole surface of the globe to the observation of civilised communities and leaves no interval of time between widely separated places proportionate to their distance apart'.[9] Standardising time became the very benchmark of the progress of civilisation. As Hallemeier, the head of the Institute for the Psychological Training of Robots, would have it in the Czech writer Karel Čapek's play *R.U.R.* (1920) (to which I shall return):

> If the time-table holds good human laws hold good; Divine laws hold good; the laws of the universe hold good; everything holds good that ought to hold good. The time-table is more significant than the gospel; more than Homer, more than the whole of Kant.[10]

A famous supporter of standard time, Count Helmuth von Moltke, appealed in 1891 to the German Parliament for its adoption, pointing out that Germany had five different time zones which impeded coordinated military planning. The insurance companies also sought uniform times to facilitate decisions on exactly when policies began and elapsed. But it was the railways which were the

---

8   *Cf.* Foucault's comments on this point in *Foucault Live*, p. 262: 'Europe was immediately sensitive to the changes in behavior that the railroads entailed.... What was going to happen when people in Germany and France might get to know one another? Would war still be possible once there were railroads? In France a theory developed that the railroads would increase familiarity among people and that the new forms of human universality made possible would render war impossible. But what the people did not foresee – although the German military command was fully aware of it, since they were much cleverer than their French counterpart – was that, on the contrary, the railroads rendered war far easier to wage.'

9   Quoted in Kern, *Culture of Time and Space*, p. 11. I also draw more generally on Kern in this discussion of time.

first to institute standard time. In 1870 there were still about eighty different railroad times in the United States alone. A single station might have clocks set to the time of two or three different meridians – one for the road east, another for west and a third for local use. On 18 November 1883, US railroads coordinated time. In 1884 representatives of twenty-five countries convened at the Prime Meridian Conference in Washington and proposed to establish Greenwich as zero meridian, to determine the exact length of the day, to divide the earth into twenty-four time zones one hour apart and to fix a precise beginning of universal time, in short to do away with the 'chaos' of the locally set clock from India to Russia to France. The wireless telegraph made it possible to maintain and transmit accurate time signals around the world, and in 1912 Paris hosted the International Conference on Time. The structure of a global system had been put in place, and in 1914 the world went to war according to mobilisation timetables facilitated by standard time.

Clearly these developments were part of a wider process, a gathering rationalisation involving new theorisations of labour, management and machinery. The introduction of time machines for workers, perfected thanks to the application of electricity, could be seen as one relatively late refinement of a tendency in operation for at least a century – the constriction and discipline of the work of the factory labourer already manifest for instance in the early nineteenth-century cotton mill:

> Yet the fact that the cotton-mills inspired such visions of working men narrowed and dehumanized altogether by completely 'self-acting' (automated) machinery is equally significant. The 'factory' with its logical flow of processes, each a specialized machine tended by a specialized 'hand', all linked together by the inhuman and constant pace of the 'engine' and the discipline of mechaniza-tion, gas-lit, iron-ribbed and smoking, *was* a revolutionary form of work.[11]

A transformation took place in the very rhythm of work – factory labour now imposed an unprecedented regularity unlike pre-industrial labour, which, for all its routine and monotony, depended on the variation of the weather and the seasons. Certainly the stress upon the horrors of modern factory conditions has led many com-mentators to romanticise the earlier rural world. Nevertheless for the artisan or the peasant there often *was* less regularised external

10 Čapek, *R.U.R.*, p. 109.
11 Hobsbawm, *Industry and Empire*, p. 68.

control over time and motion. From the early work of Marx to the sociology of Durkheim, modern mechanisation was seen as central to contemporary 'alienation' or 'anomie'. As J.A. Hobson put it in 1906:

> The defect of machinery, from the educative point of view, is its absolute conservatism. The law of machinery is a law of statical order, that everything conforms to a pattern, that present actions precisely resemble past and future actions. Now the law of human life is dynamic, requiring order not as valuable in itself, but as the condition of progress. The law of human life is that no experience, no thought or feeling is an exact copy of any other. Therefore, if you confine a man to expending his energy in trying to conform exactly to the movements of a machine, you teach him to abrogate the very principle of life. Variety is of the essence of life, and machinery is the enemy of variety.[12]

Or more recently Hobsbawm:

> Industry brings the tyranny of the clock, the pace-setting machine, and the complex and carefully-timed interaction of processes: the measurement of life not in seasons ('Michaelmas term' or 'Lent term') or even in weeks and days, but in minutes, and above all a mechanized *regularity* of work which conflicts not only with tradition, but with all the inclinations of a humanity as yet unconditioned into it.[13]

But within this broad transformation I have been trying to identify something more specific: a particular tie between the historical understanding of industrialisation and destruction, 'rationalisation' and madness. I have focused on the widely perceived shift in the 1860s; I turn later to further ramifications of this vision in the aftermath of the First World War, new tales of uncontrollable war and 'the revolt of the machines'.

In the second half of the nineteenth century notions of the army as a modern machine rather than an aristocratic 'throwback' intensified and ramified. Again there are conflicting ways one could periodise this shift. For the French philosopher and historian Michel Foucault, the time frame is large. The army was transformed in accordance with a 'disciplinary time' which marked the vast move into the modern age. The very contrast in the dramatic pages which open *Discipline and Punish* turns on the arrival of that timetable in

12   J.A. Hobson, *The Evaluation of Modern Capitalism. A Study of Machine Production*, p. 328.
13   Hobsbawm, *Industry and Empire*, pp. 85–6.

the prison. Foucault's book seeks to take us to the interface of the prison, army and factory, precisely to the crossroads where the regimentation of the body and a new precise ordering of time met: military apprenticeship was the duration of the process wherein the subject assimilated the myriad necessary gradations and manoeuvres of time and motion.[14] The aim, which Western armies have achieved with remarkably consistent success during the two hundred years in which formal military education has been carried on, is to reduce the conduct of war to a set of rules and a system of procedures – and thereby, in Keegan's words, 'to make orderly and rational what is essentially chaotic and instinctive'.[15] All the officer's professional activities are assimilated to a corporate standard and a common form. The soldier learns 'military writing' and 'voice procedure' which teach him to describe events and situations in terms of an instantly recognisable and comprehensible vocabulary, and to arrange what he has to say about them in a highly formalised sequence of 'observations', 'conclusions' and 'intentions' (pp. 18–19). As Foucault writes:

The unit – regiment, battalion, section and, later, 'division' – became a sort of machine with many parts, moving in relation to one another, in order to arrive at a configuration and to obtain a specific result. What were the reasons for this mutation? Some were economic: to make each individual useful and the training, maintenance, and arming of troops profitable; to give to each soldier, a precious unit, maximum efficiency. But these economic reasons could become determinant only with a technical transformation: the invention of the rifle: more accurate, more rapid than the musket, it gave greater value to the soldier's skill; more capable of reaching a particular target, it made it possible to exploit fire-power at an individual level; and, conversely, it turned every soldier into a possible target, requiring by the same token greater mobility; it involved therefore the disappearance of a technique of masses in favour of an art that distributed units and men along extended relatively flexible, mobile lines. Hence the need to find a whole calculated practice of individual and collective dispositions, movements of groups or isolated elements, changes of position, of movement from one disposition to another; in short, the need to invent a machinery whose principle would no longer be the mobile or immobile mass, but a geometry of divisible segments whose basic unity was the mobile soldier

14 See Foucault, *Discipline and Punish*, pp. 156–62.
15 Keegan, *The Face of Battle*, p. 18.

with his rifle; and no doubt, below the soldier himself, the minimal gestures, the elementary stages of actions, the fragments of spaces occupied or traversed.[16]

Evidently the discipline and the health of the soldier were to be linked in new ways in Foucault's modern world. But again it seems worth stressing the specificity of the later nineteenth-century period, and the 1860s in particular.

The second half of the nineteenth century witnessed a gradual consolidation of the army as a 'respectable' institution in British society, notably furthered through the Cardwell reforms. It is in and beyond the 1850s that certain modifications start to be perceptible – from the post-Waterloo image of itinerant, potentially rebellious or even revolutionary troops, to the icon of the domesticated patriotic soldier. The 'veteran' was increasingly portrayed as a symbol of the deserving poor, worthy because of his 'patriotic' service to the nation.[17] By the late nineteenth century regiments had become a source of local and civic pride. The army was 'Christianised', salvaged as the symbol of purpose, discipline and godliness. The army was to be 'saved' so it might be, truly, a salvation army for the nation. In this vision, the good army was to defend the nation abroad, just as General Booth's Salvation Army was to struggle against the depravity and the degradation of the city masses. As John Mackenzie writes, 'The cult of the Christian military hero developed out of the Indian Mutiny and reached its apotheosis in General Gordon'.[18] Both state and public schools were increasingly preparation grounds for the military. Drilling was adopted in both types of school from the 1880s. Martial activities became an important source of recreation for the working classes through the volunteer forces, rifle clubs, brigades and so forth.[19]

British public perceptions of the army as anarchic, an unruly and corrupt operation, run by snobbish, backward-looking aristocrats and manned by potentially riotous social outcasts, were gradually displaced and dissolved. New codes for drilling and moralising the soldier were elaborated; mental and physical unfitness in the army was seen as an increasingly significant scandal. The military came to

16 *Discipline and Punish*, pp. 162–3.
17 After 1815, thousands of soldiers had returned home to poverty. Only the maimed or severely diseased were entitled to a pension. The soldiers then swelled the ranks of the urban unemployed, and were to arouse powerful fears. On the shift in the iconography of the soldier during the 1850s, see J. Hichberger, 'Old soldiers'.
18 Mackenzie, *Propaganda and Empire*, p. 5.
19 See *ibid.*, p. 6.

be viewed as the litmus paper of national performance, physique and morale.[20] Boy-scout initiative was to be combined with dogged loyalty and machine-like obedience; an ethos of sport, chivalry and discipline was to be carefully cultivated in the hierarchical world of barracks and academy. This elaborately nurtured system would find its satire in the First World War.

For Foucault, this new regime of discipline, surveillance, moralisation and punishment binds together (most explicitly of all in Bentham's 'Panopticon') the genealogies of factory, barracks and prison. If we develop this broader canvas, we note (at a more prosaic level of social history than Foucault's), the interchange of personnel between prisons and armies in the nineteenth century. There was a continuing interpenetration of influences between prison and army, although the latter's penal rationale was likely to place a more explicitly utilitarian adaptive emphasis on disciplinary results than on inner religious or moral repentance in the individual.[21]

20 What was attempted ideologically was a powerful domestication of the image of the army. Perceptions of a parasitic, hard-drinking officer class ruling over ruffian privates were to be strenuously challenged and displaced. Each regiment, according to the Cardwell reforms in the 1870s, was to be based in a particular locality, its 'linked' battalions alternately serving in the Empire and at the home base. A whole series of private evangelical initiatives occurred that affected the army and navy, notably the foundation of Agnes Weston's 'Sailors' Rests'. The number of chaplains in the army was increased and they were put into uniform. Soldiers' institutes were founded and at the turn of the century, Roberts, Commander-in-Chief, founded the Army Temperance Association. See Summers, 'Militarism in Britain before the Great War'. *Cf.* Eby, *The Road to Armageddon.*

21 On the debate about the applicability of army codes to prisons in nineteenth-century Britain, see McConville, *A History of English Prison Administration,* pp. 180–1. The introduction of the prison 'silent system', for instance, was to be achieved with a deliberately 'military' efficiency and speed. It commenced with minimal preparation, literally overnight. On 29 December 1834 George Chesterton deployed his staff and issued the order whereupon: 'a population of 914 prisoners were suddenly apprised that all intercommunication by word, gesture or sign was prohibited; and, without a murmur, or the least symptom of overt opposition, the silent system became the established rule of the prison' (p. 244).

   According to William Williams, inspector for the Northern and Eastern District: 'To such as are unacquainted with the precision, regularity and minuteness, attained without difficulty under military discipline, the details of the system of silence must naturally seem cumbrous and prolix; while to myself, after some experience in the army, this does not at all appear to be the case.' He continued: 'The management of prisons should be as near as possible upon the same principles as a military organisation. The prisoners should be formed into divisions; and officers appointed to their charge; all orders and reports should be made in writing; the responsibility, duties and

In fact of course there were always discrepancies between actual prisons and their blueprints. The point is not that either prison or army functioned without friction, but that their failure or shortfall was increasingly measured against a shared vision of controlled time and disciplined motion. It was not by chance that the first three chairmen of the Directorate of Convict Prisons were officers of the Royal Engineers.[22]

The theorisation of time and motion early in the twentieth century was in many ways novel; it was not simply the end point of a longer process of rationalisation from the eighteenth century, but a response to and a catalyst of a further industrial-managerial transformation.[23] In industry the efficient recording and exploitation of time was increasingly the subject of experiment and intricate theorisation. This process would eventually be elaborated in the United States in Frederick Taylor's *Principles of Scientific Management* (1911). The creed which came to be called 'Taylorism' sought the rational scientific control of the time of the worker in an age of mass production. Taylor offered to show the manager how to eliminate unnecessary motion in the workforce; to consider the 'fitness' of each worker for his task; to undertake stop-watch observations of the efficiency of each action. The herding of men was demoralising, he argued, and reduced efficiency. The worker had to be treated as an individual, encouraged and helped by the management. Skilling, not deskilling, he enthused; industry must promote a virtuous spiral of competence, upward mobility and ambition all the way along the line.[24] Taylor proposed the industrial reconciliation of two seemingly contradictory tendencies: standardisation and individualisation. By the time of Taylor's death in 1915, the assembly line at Ford's in Detroit was, after two years of experiment, fully operational. In effect, as Peter Wollen has put

precedence of the officers should be defined; the daily routine of services laid precisely down and never departed from; the interior of a well-regulated prison should present the same aspect as a garrisoned fortress' (Quoted in McConville, *A History of English Prison Administration*, p. 245).

22   Colonel Jebb was appointed chairman upon the formation of the directorate in August 1850, where he remained until his death in 1863. He was succeeded by Colonel Edmund Henderson, who resigned in 1869. He was succeeded by Captain Edmund Du Cane (1830–1903). Given the military experience of the directors, it is not surprising that a large proportion (around two-thirds) of their staff were also drawn from army and navy backgrounds. On debates about the consequences of this link and criticism of militarism in the prison system, see *ibid.*, pp. 431, 454–5.

23   For recent discussions of these changes, see Teich and Porter (eds), *Fin-de-siècle and its Legacy*.

24   See Taylor, *The Principles of Scientific Management*, pp. 39, 41, 49, 67, 68.

it, 'Fordism turned the factory into a kind of super-machine in its own right, with both human and mechanical parts'.[25]

Various elements were involved in such new blueprints of the factory: a hierarchy of standardised, segmented and subsegmented parts, all of which were interchangeable; simultaneously a hierarchy of machine tools, also with standardised parts; a Taylorised work-force, contentedly learning and developing their skills, cooperatively suggesting improvements to management but also performing standardised repeated actions. They would be managed by an elite of engineers, supervisors and designers.

'THE VOICE OF THE MACHINES'

This new machine age was the source both of idealisation and dread. To take an extreme example of machine enthusiasm, consider Gerald Stanley Lee's *The Voice of the Machines. An Introduction to the Twentieth Century* (1906). Lee celebrated the aesthetics of the machine – the beauty of locomotives, the music of the telephone, the glory of electricity – in terms which initially seem akin to those of Italian futurism.[26] The machine, we are told, is not only a symbol of the unencumbered nature of the truly modern ('The machine has no traditions'), but the possessor of a unique poetry:

> By the side of a machine of one sort or another, whether it be of steel rods and wheels or of human beings' souls, [man] must find his place in the great whirling system of the order of mortal lives, and somewhere in the system – that is, the machine – be the ratchet, drive wheel, belt, or spindle under infinite space, ordained for him to be from the beginning of the world. (p. 7)

The machine is elevated to the sublime. Faced by the electric car, Lee felt as if he 'had seen the infinite in some near familiar, humdrum place' (p. 153). Machines are everywhere: 'We breathe the machine' (p. 14). They are a source of convenience, pleasure and ecstasy. Compared with Marinetti's, however, this is a very peaceful futurology. Lee seems uninterested in destructive machinery

25 Wollen, 'Cinema/Americanism/the robot', p. 8.
26 'Out of all the machines that [man] has made the electric machine is the most modern because it is the most spiritual. The empty and futile look of a trolley wire does not trouble the modern man. . . . The electric machine fills him with brotherhood and delight' (Lee, *The Voice of the Machines*, p. 58). *Cf.* 'ELECTRICITY – the archangel of matter' (*ibid.*, p. 59).

and war, merely offering the occasional hint that this 'sorrowing civilization of ours' has 'a kind of devilish convulsive energy in it'.

Energy was not only an elaborate cultural and political metaphor in such work as Lee's, but a star attraction of the modern fair, the object of considerable popular excitement. Take the 'exhilaration' of radioactivity, 'the fateful enthusiasm for nuclear energy that would sweep the world during the first half of the twentieth century', as Spencer Weart has described it.[27] Projections of the future city sometimes conjured up a dazzling world of science and order. On occasion there could be no doubting the popular enthusiasm for this staged modernity, this technocratic dreamland of whiteness and light.[28]

On the other hand, the vexed question and perhaps even the paradox of the relationship between science, armaments and progress haunted many pre-First World War commentators, even those who took absolutely for granted the superiority of Western civilisation over all others. The possibility of a fatal friction between technical advance and the survival of 'civilisation' became increasingly fearsome. Such a pessimistic set of perceptions was very different from the contemporaneous machine euphoria of Lee in the *Voice of the Machines* in 1906 or from numerous other commentators on both sides of the Atlantic who continued to insist on the unlikelihood or even the impossibility of modern inter-European war. But it was even more profoundly remote from the Cobdenite assumptions about peace, trade and the factory sketched out at the beginning of this study.

Of course both eulogies to the beneficence of new technology and diatribes against the effects of the 'mechanical age' had abounded in

27 Weart, *Nuclear Fear*, p. 7.
28 'In 1893 a perfected city had actually been constructed, if only for a summer. The fairgrounds of the Chicago International Exposition, dubbed the "White City", were a fairyland of broad avenues and sparkling fountains, incandescent at night under the new electric lamps with steel dynamos gleaming alongside alabaster sculptures of virgins, a picture of the future harmony between technology and art. . . . By the start of the twentieth century the image of a White City, expanded to planetary scale and projected on the blurred screen of the future, was at a peak of popularity; it was well positioned to become the first symbol associated with the energy of atoms. The connection came through another widely held idea: that modern civilization was founded, no longer on canals, but on energy. For coal and electricity were visibly transforming nations. It was a change more rapid than has happened to any generation before or since. . . . When the Museum of Natural History in New York City put a speck of radium on display in 1903, one of the largest crowds the museum had ever admitted swarmed in, squeezing and elbowing, to stare at the dull pinch of powder' (*ibid.*, pp. 7–10).

the first half of the nineteenth century too; certainly the advent of new machines was powerfully contested in that earlier period, but primarily in terms of its social cost rather than military uncontrollability. Between Carlyle or Cobden on the one side, and the anxious interrogation of technology in the period before 1914 on the other, lay a dramatic transformation, the decisive experience of the Franco-Prussian War, radically different material conditions of conflict. Cobden, as I indicated, touches on this novelty in the last decade of his life, the 1860s, but his earlier model of war and peace remained largely intact.

The American Civil War ushered in an age in which characteristically major warfare would be endured by populations as a whole rather than professional armies and those unfortunate enough to lie in their path. By the 1860s the issue of the industrial revolution was coming to be perceived as central to the question of war, even if for a further period chivalric and Napoleonic war codes and images still persisted alongside.[29] Here was a new situation of total war which could be contrasted with even the most terrible precedents. The Civil War was to witness a devastating increase in firepower, which in turn ushered in the entrenched battlefield.[30] The Civil War soldier was obliged to dig in on the attack as well as on the defence – thus anticipating the First World War. Moreover, allowing for British experiments in the Crimea, the electric telegraph now made its first substantial entrance into battle, as did an advanced system of visual signals that used the telegraphic alphabet. Again despite limited precedents, this period was the first to see the significant technological development and application of gas-filled observation balloons in warfare – and hence of military aeronautics. Importantly, the American Civil War armies were the first to use railroads to move and supply troops in the field. The great expanses of the American battlefield meant the collapse of the traditional transportation and supply standards of Western European armies.[31] It was one mark of a wider shift in the cultural conception of time and space discussed above.

If the American Civil War, as is often argued, had a relatively limited influence on European military thought,[32] it would be dif-

---

29 As Michael Howard puts it: 'War was no longer a gladiatorial conquest between professional forces. It had not been so since the middle of the nineteenth century, but it took a good sixty years for the lesson to sink in'; *Studies in War and Peace*, p. 21.
30 See Hagerman, *American Civil War*, chs 7 and 8.
31 See *ibid.*, pp. xi–xiii.
32 See most recently *ibid.*, *passim*.

ficult to overestimate the impact of the Franco-Prussian War in
1870. For many late nineteenth-century commentators, 1870 constituted the disturbing threshold of modernity. Whilst it is not the
aim of the present study to downplay the impact of the First World
War, it is also important to see the extent to which representation
after 1914, including the representation of the First World War as
utterly novel, drew on earlier perceptions and anticipations of the
'modern' war. There were many continuities between the pre- and
post-1914 representation of war. Neither trenches nor shellshock,
for example, were simply 'invented' in 1914.[33]

THE PERFECT ABATTOIR

In the 1860s, technology, factory production and calculated death
were coming together in many new ways. Take the modern
assembly-line slaughterhouse. Just as rails now transported soldiers
to battle, they bore animals to their death. The coming of the
railway in the first half of the nineteenth century had opened up
new possibilities in the assembly process, enabling the reduction of
delays and 'friction' between each productive stage.[34] Assembly
lines had developed for textiles and food during the 1830s and
1840s. Rails were adapted in a variety of ways to the factory;
tracks, trolleys, belts and cranes all saw considerable innovation
and development. Overhead rail systems appeared first on a large
scale in American slaughterhouses of the Middle West during the
late 1860s.[35]

It was Baron George Eugène Haussmann, architect of Second
Empire Paris, who projected most dramatically the European vision

33 On shellshock in the American Civil War, see Veith, *Hysteria*, p. 212ff;
    Benison *et al.*, 'Walter B. Cannon and the mystery of shock', p. 219. The
    Civil War was to see the use of military trenches, railways, vast and moving
    theatres of battle, mass mobilisation and civil destruction as war strategy
    rather than by-product.

34 Between Adam Smith's *An Enquiry into the Nature and Causes of the Wealth
    of Nations* (1776), and Frederick Taylor's *Principles of Scientific Manage-
    ment*, on the eve of the First World War, lay a range of studies aiming to
    investigate and rationalise each productive operation. Mention must be made
    of Charles Babbage, Professor of Mathematics at Cambridge who sought to
    cost and time each operation in needle manufacture. See Babbage, *On the
    Economy of Machinery and Manufacture* (1832); *cf.* Giedion, *Mechanization
    takes Command*, p. 114.

35 See Giedion, *Mechanization takes Command*, p. 78.

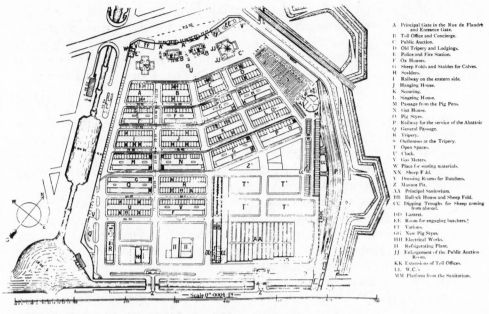

A Principal Gate in the Rue de Flandre
   and Entrance Gate.
B Toll Office and Concierge.
C Public Auction.
D Old Tripery and Lodgings.
E Police and Fire Station.
F Ox Houses.
G Sheep Folds and Stables for Calves.
H Scalders.
I Railway on the eastern side.
J Hanging House.
K Scouring.
L Singeing House.
M Passage from the Pig Pens.
N Gut House.
O Pig Styes.
P Railway for the service of the Abattoir
Q General Passage.
R Tripery.
S Outhouses at the Tripery.
T Open Spaces.
U Clock.
V Gas Meters.
W Place for storing materials.
XX Sheep Fold.
Y Dressing Rooms for Butchers.
Z Manure Pit.
AA Principal Sanitorium.
BB Bullock House and Sheep Fold.
CC Dipping Troughs for Sheep coming
   from abroad.
DD Lazaret.
EE Room for engaging butchers.
FF Various.
GG New Pig Styes.
HH Electrical Works.
I Refrigerating Plant.
JJ Enlargement of the Public Auction
   Room.
KK Extensions of Toll Offices.
LL W.C.'s
MM Platform from the Sanitorium.

7   Abattoirs of La Villette, Paris (from *Douglas's Encyclopaedia*).

8   Live Stock Markets adjoining Abattoirs of La Villette, Paris (from *Douglas's Encyclopaedia*).

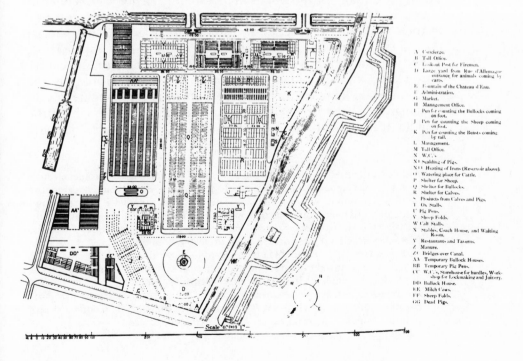

A Concierge.
B Toll Office.
C Look-out Post for Firemen.
D Large yard from Rue d'Allemagne
   entrance for animals coming by
   carts.
E Fountain of the Chateau d'Eau.
F Administration.
G Market.
H Management Office.
I Pen for counting the Bullocks coming
   on foot.
J Pen for counting the Sheep coming
   on foot.
K Pen for counting the Beasts coming
   by rail.
L Management.
M Toll Office.
N W.C.'s
NΙ Scalding of Pigs.
NΙΙ Heating of Irons (Reservoir above).
O Watering place for Cattle.
P Shelter for Sheep.
Q Shelter for Bullocks.
R Shelter for Calves.
S Products from Calves and Pigs.
T Ox Stalls.
U Pig Pens.
V Sheep Folds.
W Calf Stalls.
X Stables, Coach House, and Waiting
   Room.
Y Restaurants and Taverns.
Z Manure.
ΖΙ Bridges over Canal.
AA Temporary Bullock Houses.
BB Temporary Pig Pens.
CC W.C.'s, Storehouse for hurdles, Work-
   shop for Lockmaking and Joinery.
DD Bullock House.
EE Milch Cows.
FF Sheep Folds.
GG Dead Pigs.

of the ideal slaughterhouse of the future.[36] In question was the possibility of a perfectly engineered, centralised, hygienic meat location, catering to the needs of millions. Under the superintendence of the architect M. Janvier, Haussmann allowed the building of the central slaughterhouse to go ahead in 1863; it opened in 1867 at La Villette, although it was still not complete when the Second Empire collapsed. Remarkable for its planning and its scale, La Villette retained elements of a handicraft tradition at odds with the routinised mass slaughter to come later. Aspects of the killing arrangements remained cumbersome, since each ox was held in a separate booth. With the development of the Union Stockyards of Chicago in the 1860s, however, mechanised animal butchery moved towards its apotheosis. Chicago became the greatest cattle market in the world and quickly reached the point of slaughtering and processing some 200,000 hogs a day. Carcasses could now be transported over vast distances; a series of experiments during the 1870s with refrigeration enabled meat to be sent across the United States and even abroad. In 1876 the Frenchman Charles Tellier had succeeded in bringing meat across the ocean in the ship *Frigorifique*. From the 1870s to the 1880s a series of refinements occurred in the apparatus used for killing and processing the hogs.[37]

In their specialised division of labour, the new slaughterhouses exemplified many of the points elaborated in Taylor's *Principles of Scientific Management*. The abattoir was to become a complex factory involving a precise separation of tasks and mechanical operations – stabling, killing, cleaning, refrigerating, transporting, inspecting, preparing foodstuffs, and so forth. In nineteenth-century Europe the killing gradually became less socially visible, as the function of the butcher was separated off from the slaughterer and as slaughterhouses were relocated outside the city. Throughout the Middle Ages and until the eighteenth century, the Grande Boucherie

36 Although the need to eliminate old and small abattoirs was a problem which had already preoccupied Napoleon I. In 1807 a decree had ordered the building of public slaughterhouses. The first was completed in 1810 and Napoleon ordained that abattoirs were to be established throughout France; see Giedion, *Mechanization takes Command*, p. 209; *Douglas's Encyclopaedia*, p. 1. Nevertheless Haussmann's later project was unprecedented in its scale and its attention to detail: 'Haussmann toiled over the slaughterhouse of La Villette with painstaking care, one might almost say with the consciousness of a mission to fulfil.... It became *the* abattoir, a prototype for the rest of the century, just as the boulevards and public parks of Haussmann's Paris became models from which every growing metropolis of the Continent took pattern' (Giedion, *Mechanization takes Command*, p. 210).

37 Giedion, *Mechanization takes Command*, pp. 221, 231–2.

de Paris, for instance, had been located in the centre of the capital.[38]
But with the rise of new conceptions of hygiene, smell, squalor and
city planning, abattoirs were to be moved away from crowded
centres. New codes of hygiene came into operation in the abattoir
to avoid contamination – wood for instance was no longer to
be allowed on knife handles, as it was a material susceptible to
germs. Requirements of dress, equipment cleaning and treatment of
animals were introduced, not least to seek to prevent the passing of
tuberculosis via animal flesh. It was also forbidden to suspend an
animal before it was insensible.

More piecemeal measures were afoot in Britain too, although
reformers continued to lament the lack of a continental system
of state regulation and central uniform inspection well into the
twentieth century. 'Not only are live animals driven or carted
through the streets to the slaughter-houses, but carcasses are conveyed in open vans, and blood and refuse pass through the sewers
of the greatest city of the world,' complained R. Stephen Ayling of
the Royal Institute of British Architects in 1908.[39] America, France
and Germany were cited as beacons of the future, putting England's
meat industry to shame.[40] Architectural schemes were carefully
worked out for the perfectly rational slaughterhouse – with the
precise compartmentalisation of lairs, slaughter halls, cooling

38 The Grande Boucherie was in the area now known as Châtelet les Halles. It
    was close to the fish market, the pillory and the Place de Grève (where
    executions took place). See Vialles, *Le Sang et la chair*, p. 15.

39 Ayling, *Public Abattoirs*, p. 5. Ayling lamented the poor hygiene, the
    unscrupulous private profiteering which triumphed over public health,
    the inhumanity of the treatment of animals, the poor moral status of the
    slaughterman and the pathetic failure to impose a standard system, despite
    numerous government inquiries around the turn of the century, not least the
    Royal Commission of 1896. By contrast, he declared, 'In 1897 I spent a
    month in the abattoirs of La Villette, Paris, and during that time, I am
    pleased to say, I never saw an animal kicked, its tail twisted, or any unneces-
    sary force used either in the cattle market or slaughter-houses' (pp. 10–11).
    The modern slaughterhouse needed constant inspection, regulation, sanitisa-
    tion and moralisation: 'it should be made quite possible for any ratepayer,
    or indeed any person interested in the subject, to obtain an order and be
    conducted over the buildings by a competent guide' (pp. 13–14). They
    should be publicly run, but self-supporting, indeed profitable. All parts of the
    animal should be carefully husbanded (for manufacture of dyes, albumen,
    manure and so forth) and not allowed to drift into the public sewers.

40 See *Douglas's Encyclopaedia*, pp. 1–2; Schwarz, *Public Abattoirs*; Cash and
    Heiss, *Our Slaughter House System. A Plea for Reform, and The German
    Abattoir*. The favourable view of the American slaughterhouse industry would
    have received something of a jolt, to say the least, from a reading of Sinclair's
    The Jungle.

9 Interior of pig carcase hanging hall, La Villette, Paris (from R. Stephen Ayling, *Public Abattoirs: Their Planning, Design and Equipment*, 1905).

chambers, cold storage and chill areas, slaughter rooms for infected animals, offal stores, workmen's messroom, kitchen, baths and lavatories.[41]

The most famous fictional evocation of the slaughterhouse is surely Upton Sinclair's socialist novel, *The Jungle* (1906), which recounts in detail the social and industrial world of the Chicago stockyards as encountered by the newly arriving East European immigrant, Jurgis Rudkus, and his extended family. The killing factory with its 'stream of animals' employs and then consumes the newcomers: 'it was quite uncanny to watch them, pressing on to their fate, all unsuspicious – a very river of death.'[42] For Sinclair the abattoir becomes a vast metaphor of human destiny under capitalism. 'It was the Great Butcher – it was the spirit of Capitalism made flesh' (p. 377). The ghastly description of this 'slaughtering-machine' runs on for pages and pages, detailing the agonised shrieks of the animals, the blood and guts, the unspeakable horror of it all. It vividly combines a sense of extraordinary technological advance

41 See Ayling, *Public Abattoirs*, p. 32 and Figs 7–11.
42 Sinclair, *The Jungle*, p. 42.

10 'The Old Shambles – Leeds Slaughter-house now superseded' (from *Douglas's Encyclopaedia*).

11 'Leeds City Meat Market and Slaughter-houses, interior view of Market' (from *Douglas's Encyclopaedia*).

with grotesquely primitive working conditions, profound human degradation, corruption and graft. Thus the 'wonderful machine with numerous scrapers, wheels and cogs and belts' ('pork-making by applied mathematics') is situated amidst the extreme squalor of the poor workers' domestic and factory existence.

Sinclair meticulously recounts the course of the pig through the processing machinery; and amidst it all he details the human alienation of the factory operatives: 'Looking down this room, one saw, creeping slowly, a line of dangling hogs a hundred yards in length; and for every yard there was a man, working as if a demon were after him' (p. 46). Everybody has their post within the division of labour. The individual must accomplish his or her specific action at a frenzied pace, often amidst a pool of blood. Every part of the animal body is processed into a commodity: grease becomes soap and lard; bristles become the hair for cushions; the skins are dried and tanned; the heads and feet are made into glue; the bones into fertilizer. The horns of the cattle are transformed into combs, buttons, hairpins and imitation ivory; shin bones become toothbrush handles and mouthpieces for pipes; the hoofs become hairpins and buttons; the feet, knuckles, hide clippings and sinews become gelatine, phosphorus, bone black, shoe blacking and bone oil. Blood is processed into albumen; entrails emerge as violin strings, and so forth (pp. 50–1). 'The killing beds' form an industry on a vast scale. It is a *laissez-faire* marketplace of human labour, at times evoked in Darwinian (or perhaps Spencerian), and at times in Hobbesian life sentences: 'It was a war of each against all, and the devil take the hindmost' (p. 91).

If the vision of the slaughterhouse is linked throughout to the nature of the capitalist free market, only close to the end of the novel are we offered an explicit connection to the world of modern war, that other slaughterhouse scene. Jurgis finally discovers socialism. He is enraptured by the speaker at a political meeting who makes the association for him between capitalism, slaughter and war:

> Realize that out upon the plains of Manchuria tonight two hostile armies are facing each other – that now, while we are seated here, a million human beings may be hurled at each other's throats, striving with the fury of maniacs to tear each other to pieces! ... We call it War, and pass it by; but do not put me off with platitudes and conventions. Come with me, come with me – *realize it*! See the bodies of men pierced by bullets, blown into pieces by bursting shells! Hear the crunching of the bayonet, plunged into human flesh; hear the groans and shrieks of agony,

see the faces of men crazed by pain turned into fiends by fury and hate! Put your hand upon that piece of flesh – it is hot and quivering; just now it was a part of a man! This blood is still steaming – it was driven by a human heart! Almighty God! And this goes on – it is systematic, organized, premeditated! (p. 362).

The 1860s was to inaugurate a new systematic mechanisation of death in both military and industrial killing machinery, albeit with uneven consequences in different countries. A new 'humane' order of killing (the world of the Hague conferences applied to animal slaughter, as it were) went hand in hand with a new vastness of death, a hitherto inconceivably rationalised and industrialised processing of meat. Not by chance was the metaphor of the slaughterhouse to become so inextricably intertwined with the language of modern war; they emerged so closely together. As Marx wrote to Engels in 1866: 'Is there anywhere where our theory that the *organisation of labour is determined by the means of production* is more brilliantly confirmed than in the human slaughter industry?'[43] The interconnecting resonances seem unavoidable in that question, as again they do, for example, in this graphic account of the slaughterhouse, written just beyond 1945 (which, as the author makes plain, adds to the passage a further horrible historical connotation):

The death cries of the animals whose jugular veins have been opened are confused with the rumbling of the great drum, the whirring of gears, and the shrilling sound of steam. Death cries and mechanical noises are almost impossible to disentangle. Neither can the eye quite take in what it sees. The one side of the sticker are the living; on the other side, the slaughtered. Each animal hangs head downwards at the same regular interval, except that, from the creatures to his right, blood is spurting out of the neck-wound in the tempo of the heart beat. In twenty seconds, on the average, a hog is supposed to have bled to death. It happens so quickly, and is so smoothly a part of the production process, that emotion is barely stirred.

What is truly startling in this mass transition from life to death is the complete neutrality of the act. One does not experience, one does not feel; one merely observes. It may be that nerves that we do not control rebel somewhere in the subconscious. Days

43 Marx and Engels, *Correspondence. A Selection, 1846–95*, p. 209.

later, the inhaled odor of blood suddenly rises from the walls of one's stomach, although no trace of it can have clung to the person.

How far the question is justified we do not know, nevertheless it may be asked: Has this neutrality toward death had any further effect upon us? This broader influence does not have to appear in the land that evolved mechanized killing, or even at the time the methods came about. This neutrality towards death may be lodged deep in the roots of our time. It did not bare itself on a large scale until the War, when whole populations, as defenceless as the animals hooked head downwards on the travelling chain, were obliterated with trained neutrality.[44]

In a lucid study, finely setting out and extending many of the historiographical themes I have raised here, Zygmunt Bauman has recently repudiated the idea that the Holocaust can be confined to a history either of Jewish victims or of German perpetrators:

The implication that the perpetrators of the Holocaust were a wound or a malady of our civilization – rather than its horrifying, yet legitimate product – results not only in the moral comfort of self-exculpation, but also in the dire threat of moral and political disarmament. It all happened 'out there' – in another time, another country.[45]

To constrain the terms of the Holocaust in that way is even to participate in its logic: 'Like the Holocaust itself, its causes were enclosed in a confined space and a limited (now, fortunately, finished) time' (p. xii). Bauman's thesis is that the lessons of the Holocaust must be assimilated as part of the mainstream of our theory of modernity. The interaction of certain old tensions that modernity ignored, slighted or failed to resolve with the powerful technology and rationality of the industrial world produced a quite unprecedented 'murdering machine' (p. 23). This is not a deterministic argument in which industrialisation is seen as inevitably leading to that form of genocide, but it is a recognition of certain necessary preconditions: 'Modern civilization was not the Holocaust's *sufficient* condition; it was, however, most certainly its *necessary* condition' (p. 13). The modern state with its desire and capacity to monopolise the means of violence, must be given a central place within our understanding of what happened under

---

44 Giedion, *Mechanization takes Command*, p. 246.
45 Bauman, *Modernity and the Holocaust*, p. xii.

Nazism: the camps were state-run death factories. They cannot adequately be seen as simply the apogee of some ancient barbarism or the culmination of some timeless anti-Semitism. Drawing on Claude Lanzmann's film, *Shoah*, Bauman insists on the crucial role of petty-officialdom in the Holocaust, the cogs and wheels, the clerks and drivers, the sheer enormity and meticulousness of the whole logistical operation. In this modern murder machine, death was increasingly screened off from its perpetrators. The vast bureaucracy needed for this task was, for the most part, spared the sight of death – a disjunction which was to prove incomparably advantageous to the efficiency of the process. The victims, who had already ideologically been stripped of their humanity (pushed outside the 'universe of obligation'), were to die, as far as possible, unseen – which is not to say that those who did not see did not know: not seeing was an active process, an effect moreover of policy, not merely some 'passive' and 'neutral' condition or quirk of fate.

Whilst the investigation or interpretation of the Holocaust remains beyond the scope of my study, I cite Bauman's book at some length here because it sets out with particular clarity a dilemma about the consequences of industrialisation and modernisation, which has, in a broader sense, been at issue through much of my material and discussion. The author charts the ways in which mass murder came to be authorised, routinised, dehumanised; the way in which the Final Solution was bureaucratised and mechanised. Thus Ohlendorf, Commander of the *Einsatzgruppe*, could declare at Nuremberg: 'I surrender my moral conscience to the fact I was a soldier, and therefore a cog in a relatively low position of a great machine'.[46] As Bauman writes:

> [Auschwitz] was also a mundane extension of the modern factory system. Rather than producing goods, the raw material was human beings and the end product was death, so many units per day marked carefully on the manager's production charts. The chimneys, the very symbol of the modern factory system, poured forth acrid smoke produced by burning human flesh. The brilliantly organized railroad grid of modern Europe carried a new kind of raw material to the factories. (p. 8)

Well before the Holocaust, and often within frames of reference remote from the question of anti-Semitism, allusions abound to this emerging nexus of factory and death, a shared currency of language

46   Quoted in *ibid.*, p. 22.

and experience which I have tried to bring forth in the present chapter. From Sinclair's *Jungle* to Eisenstein's *Strike*,[47] to take two powerful examples, we confront this conjuncture, this shocking 'cut' between the slaughter of 'livestock' and people, between the methodical extermination of animal and human life.

47 *Strike* was premiered in 1925. My point does not depend upon direct 'influence' from one work to the other; on whether Eisenstein directly knew of *The Jungle* at that time; the issue is not the timing of Sinclair's (unhappy) collaboration with the Soviet film-maker, which in fact occurred only later, on *Que Viva Mexico*, 1931; see Eisenstein, *The Complete Films*, pp. 28, 79. My suggestion here is that, whether knowingly or not, we find a shared set of figurative links; across different work, from poetic anthem to cinema montage, the imaginative join between industrial meat processing and human killing – 'those who die as cattle' (Owen's 'Anthem for Doomed Youth') – seemed so powerfully right.

13 'The Unnatural and Terrible Wall of the War'

The vexed question of the First World War as archetype of the modern war (and world) demands further analysis. 1914–18 is distinctive both in its scale – its sheer labour-intensiveness – and its capacity to amalgamate so many different forms of destruction. Arguably it is just this heterogeneity of forms of destruction, sabotage and mass death rather than some new monolithic experience of violence which characterised the First World War. But that insistence on the multiple levels of social, economic, cultural and political impact still rightly leaves open the issue of discursive continuity – the contexts of language and understanding within which the difference of the war could be registered. The question remains to what extent we need to look back before 1914 to assemble the textual figures of 'the Great War' and in particular the complex contradictory resonances of a double analogy: of military conflict as anarchy and madness, and of industrialised war as the symbol of modernity. A central area of investigation in this work has been the image of war as the unstoppable and all-consuming machine, the runaway vehicle, the horse that rides its rider; of war even as the unconscious of progress; the driving force which does not recognise the word 'no' and which, ultimately, cannot be fully repressed; which breaks through illusory order, the surface world of 'progress' or 'decadence' and its attendant 'diplomacies'; which slips through in the interstices of everyday national life. Such images, of course, are not the exclusive preserve of the post-1914 world. Moreover, if it is either 'totality' or 'variety' which defines the modernity of warfare then we are still talking about relative differences. The notion and the periodisation of a founding modern war remains open to debate.[1]

1   In the words of one recent commentator: 'the French Revolutionary wars did constitute the first modern war, not because it represented a conflict between two diametrically opposed ideologies, but because of what it became. By liberating their states in the course of 1792–94, by casting off all inhibitions and maximising their resources, the French revolutionaries waged total war of unprecedented intensity and on an unprecedented scale. Not only did it 'revolutionise the Revolution' it also forced the other European states into emulation, however delayed and partial. After almost a quarter of a century of devastation, exploitation and over-exertion, no part of Europe was untouched: politically, economically, socially, intellectually or culturally. It was not the French Revolution which created the modern world, it was the French

For so many soldiers in the First World War, as Eric Leed has shown, the very meaning of the world itself was seen to be utterly and hopelessly transformed by the conflict.[2] Wartime writers often understood their contemporary history as irrevocably split from the past. Modernity was born in an absolute schism and yet at the same time was often grasped as tragic culmination of the anterior system. Past and present were organically connected; yet separated by a deadly abyss. Take, as examples, two remarkable wartime letter-writers. First, the poet Rainer Maria Rilke.

To read Rilke's wartime correspondence is at once to confront the motif of 1914 as division. The First World War marks an absolute experiential divide, an historical barrier, a linguistic limit. It is Rilke who speaks of 'the unnatural and terrible wall of the war'.[3] War shatters conventional terms of periodisation and disorientates the sequence of time. 'The past remains behind, the future hesitates, the present is without foundation.'[4] Space and time – the 'weight of the time' that now 'lies upon the slightest communication and expression'[5] – need radical reappraisal. He speaks of 'the confused interruption'[6] and 'the disastrous breaking off of a former world'.[7] Modernity has convulsed time itself; the pre-war calendar ceases to have any meaning: 'The new year was no perceptible division at all for me either, since after all the year of war in which we are living continues and cancels all other calendars, far into the season.'[8] Or again:

> A few days later fate broke out of the foggy world, this world-noise which overnight drowned out one's thinking back and on beyond what no one is yet able to think. Not even now when so much dreadful accomplishment and progress has been going on all sides. . . . To me it is an unspeakable suffering and for weeks I have been understanding and envying those who died before it, that they have no longer to experience it from here; for some-where in space there will surely be places from which this mon-strosity still appears natural, as one of the rhythmic convulsions

Revolutionary wars' (Blanning, *Origins of the French Revolutionary Wars*, p. 211).

2   See Leed, *No Man's Land*.
3   Rilke, *War Time Letters*, p. 86: 19 Nov. 1917.
4   *Ibid.*, p. 22: 6 Nov. 1914.
5   *Ibid.*, p. 25: 8 Feb. 1915.
6   *Ibid.*, p. 90: 3 July 1918.
7   *Ibid.*
8   *Ibid.*, p. 23: 4 Jan. 1915.
9   *Ibid.*, p. 20: 21 Oct. 1914.

of the universe which is assured in its existence, even where we go under.[9]

What the war offers is a new vantage point on the world that preceded it, the schism between one history and another: 'The war could be nothing other than ending; it was an extreme, following its own inner anomalousness, a breaking-off of humanity from itself. Only a new beginning of existence could set in after it.'[10] This was not to say that the war offered progress or transformation, or even some quantitative increase in suffering; only a certain crisis of realisation:

Nor is more to be achieved now, than the soul's endurance, and misery and evil are perhaps not any more present than before, only more graspable, more active, more apparent. For the misery in which mankind has daily lived since the beginning is really not to be increased through any circumstances. But increases of insight there surely are in the unspeakable misery of being man and perhaps all this is leading to them; so much decline – as though new ascents sought – distance and space for their running off.[11]

The war involves the reconceptualising of time, the recognition of ruptures and the subtle retracing of the immanence of the future in the past. For against the image of 1914 as impermeable prison wall there is the image of 1914 as a kind of porous membrane through which something trickles back, 1914–18 reshaping the memory and the historical reality of the pre-war world: 'Was it that? I say to myself a hundred times, was it that, which in these last years has been lying upon us as a monstrous pressure, this frightful future that now constitutes our cruel present?'[12] Or again:

Doesn't one seem, I said to myself, to be moving in a world through whose greedy fingers, its best inheritances have already been slipping for decades: for it is perhaps something almost imperceptible that gives all heritage its significance ... I long for people through whom the past in its large lines continues to be connected with us, related to us; for how much the future,

10 Ibid., p. 120: 1 March 1919.
11 Ibid., p. 22: 6 Nov. 1914.
12 Ibid., p. 38: 2 Aug. 1915. One might compare the following from All Quiet on the Western Front: 'In the quiet hours when the puzzling reflection of former days, like a blurred mirror, projects beyond me the figure of my present existence, I often sit over against myself, as before a stranger, and wonder how the unnameable active principle that calls itself Life has adapted itself even to this form' (Remarque, All Quiet, p. 296).

particularly now – the bolder and more daring one imagines it –
is nevertheless still going to depend on whether it falls in with the
direction of the deepest traditions and moves and is projected out
of them (and not out of negation).[13]

Rilke weighs the war in terms of suffering's density and distribution. On the one hand the war changes nothing since at any given
time, 'The whole of distress is always in use among men, all there
is of it, a constant, as there is also a constant of happiness; only the
distribution varies'.[14] But on the other, 'the appalling calamity
creates a new scale of sensation, since it reaches so deep down, it
also rises higher; is it more too, what we feel? Or do we simply read
life's degrees in Fahrenheit instead of the usual Réamur?'[15]

Rilke pores over the relationship between past and present during
the First World War, the question of tradition, of rupture, of
human continuity, of history, of distortion: 'through the present
malicious confusion, I have come to be so suspicious of everything
human, even far into the past'.[16]

The war at once clarifies meanings and dissipates sensations,
hence 'my indescribably benumbed and inhibited state of mind';
'this condition of inwardly turning to ice'; 'my heart almost in-
accessible to me'; 'I am continually shut out from everything'.[17]
The war is both a culmination of suffering and an evacuation of
feeling:

> for me, all that, all that is biggest and stirring, remains attached
> to the *other* world, the earlier, the former world, in which I
> had long been a sufferer, but never a numb person, never an
> emptied-out person, never a person shouted at who does not
> understand.[18]

Rilke's wartime letters provide just one instance among many of
how preceding nineteenth-century peace was now understood in
two opposing ways: as a world of lost innocence; and as a state of
latent war. As Kant had observed long before in *Perpetual Peace. A
Philosophical Sketch* (1795), since peace means more than a truce,
namely an end to all hostilities, 'to attach the adjective "perpetual"
to it is already suspiciously close to pleonasm'.[19] To suggest that
the addition of the word 'perpetual' to the word 'peace' is redun-

13   Rilke, *War Time Letters*, pp. 95–6: 22 Sept. 1918.
14   *Ibid.*, p. 29: 28 June 1915.
15   *Ibid.*, p. 29: 28 June 1915.
16   *Ibid.*, p. 95: 22 Sept. 1918.
17   *Ibid.*, p. 97: 9 Oct. 1918.
18   *Ibid.*, p. 34: 12 July 1915.
19   Kant, *Perpetual Peace*, in *Political Writings*, p. 93.

dant since peace can only mean something permanent – an end to war – also implies that a war undoes the illusion of the 'peace' which precedes it. War is the crisis not only of the present, but of the past. The 'pre-war' is now precisely that – a history to be read teleologically, a history tending towards overt world war, making explicit what was latent.

Now consider Henry James. Writing the day after the British entered the war in 1914, he made the well-known remark:

> The plunge of civilization into this abyss of blood and darkness by the wanton feat of these two infamous autocrats is a thing that so gives away the whole long age during which we have supposed the world to be, with whatever abatement, gradually bettering, that to have to take it all now for what the treacherous years were all the while really making for and *meaning* is too tragic for any words.[20]

James writes of the silence, the separation, the mountain, the block, which lies between August 1914 and the pre-war world. Modernity is now a 'nightmare of the deepest dye', 'a huge horror of blackness'.[21] It is afflicted by abnormal convulsions, meaningless rush and confusion brought about by the deep German and Austrian conspiracy for violence. James aims to retain 'a little civilization' in the face of the abyss, whilst also acknowledging that writing finds itself tainted by this 'sickening blackness':[22]

> My aged nerves can scarcely stand it, and I bear up but as I can. I dip my nose, or try to, into the inkpot as often as I can; but it's as if there were no ink there, and I take it out smelling gunpowder, smelling blood, as hard as it did before.[23]

The outbreak of war in 1914 utterly extinguishes the old world,[24] but, as in Rilke, it is the retrospective knowledge of the immanence of the catastrophe in the pre-war era which is unbearable. War offers a retrospective knowledge of the latent forces shaping, but repressed from, the consciousness of pre-war culture:

> The tide that bore us along was all the while moving to *this* as its grand Niagara – yet what a blessing we didn't know it. It seems to me to *undo* everything, everything that was ours, in the most

20 *The Letters of Henry James*, vol. II, p. 398.
21 *Ibid.*: 4 Aug. 1914.
22 *Ibid.*, p. 400: 6 Aug. 1914.
23 *Ibid.*, p. 402: 8 Aug. 1914.
24 'One of the effects of this colossal convulsion is that all connection with everything of every kind that has gone before seems to have broken short off in a night...' (*ibid.*, pp. 416–17: 2 Sept. 1914).

horrible retroactive way – but I avert my face from the monstrous scene![25]

The simultaneous sense of disjunction from and connection with history before 1914 is matched spatially by James's disorientated sense of geographical proximity to and remoteness from the war. From the English countryside, the season remains 'monstrously magnificent', 'and we look inconceivably off across the blue channel, the lovely rim, towards the nearness of the horrors that are in preparation just beyond'. There is something inconceivable about the nearness of the carnage that lies 'over the blue channel of all these amazing days, toward the unthinkable horror of its almost other edge'.[26]

Further examples of this move between 'schism' and 'latency' theory can be multiplied. Thus the British biologist, social reformer and pioneer in urban planning Patrick Geddes, commenting on the war in the *Sociological Review* in 1915:

It will not be denied that the peace of the past generation, especially since 1870–71, has been no peace, but one of latent war. So plainly, so fully, has this been the case, that there are many to whom the extreme state of war preparations has seemed, if not the very norm of human existence, at any rate its inevitable burden. . . . Grant by all means, that we in our lifetime have practically only known wars and rumours and preparations of wars – that, when not in patent war, we have lived in latent war.[27]

Or consider Freud in his 1915 paper 'On transience':

the war broke out and robbed the world of its beauties. It destroyed not only the beauty of the countrysides through which it passed and the works of art which it met with on its path but it also shattered our pride in the achievements of our civilization, our admiration for many philosophers and artists and our hopes of a final triumph over the differences between nations and races. It tarnished the lofty impartiality of our science, it revealed our instincts in all their nakedness and let loose the evil spirits within us which we thought had been tamed for ever by centuries of continuous education by the noblest minds. It made our country small again and made the rest of the world far remote. It robbed

25 *Ibid.*, p. 403: 10 Aug. 1914.
26 *Ibid.*, p. 405: 19 Aug. 1914; p. 407: 22 Aug. 1914.
27 Geddes, 'Wardom and peacedom', p. 20.

us of very much that we had loved, and showed us how ephemeral were many things that we had regarded as changeless.[28]

The First World War, then, was not only counterposed to earlier 'innocence', but was also understood as *revelation* of the folds of denial within its own past. The nineteenth century *became* the gathering reservoir of the war, which was perceived as the definitive perspective on the complex many-layered world of apparent peace which had preceded it, a peace which had actually contained many 'small' wars within it. As Dr Arthur Brock was to suggest in a 1918 discussion of shellshock:

> We have seen in the war neurasthenic an obsession implanted by the frightfulness of the battlefield and the spectre of the blood-stained Boche. But are not these horrors of war the last and culminating terms in a series that begins in the infernos of our industrial cities? Think of the mental anguish inflicted on families subjected to the struggle-for-life in these torture chambers of our competitive world during that recent phase of a 'peace' which we now see to have been but latent war.[29]

And as he continued: 'Each of these shell-shock hospitals can be looked on as a microcosm of the modern world, showing the salient features of our society (and especially its weaknesses) intensified, and on a narrower stage' (p. 25).

For Rilke, James, Freud and numerous other contemporaries, 1914 represented the inescapably telling landmark of the modern age. They evoke a much wider motif in twentieth-century culture and historiography where 1914 is cast as both conclusion and as origin, outlet and source of the Nile of modern experience and derangement. On the one hand pre-1914 society was cast as innocent, vacuous, empty of the shattering experience of world war; on the other hand as a kind of latency period in a longer process of development and confrontation.

There is now an extensive secondary literature on late nineteenth-century militarism, the arms race, the navy race, imperialist rivalry, shifting alliances and antagonisms; historians make reference to uncanny pre-war anticipations (Bebel, Bloch, Engels, Wells) and yet it is also an historiographical commonplace to locate the First World War itself as the definitive birthplace of modern language and meaning, the origin of 'the contemporary world', the twentieth-century period that is so often dated from 1914.

---

28 'On transience' (written in 1915, published 1916), p. 307.
29 Brock, 'The re-education of the adult', p. 40.

The move from pre-war to First World War is thus portrayed as a simple passage from innocence to experience, too easily perpetuating the nineteenth-century Hegelian or social Darwinist myth of war as a decisive (unidirectional) forward or backward movement in the psychological, racial or organic 'life' of a culture or nation. The military conflict is sometimes seen as a sweeping away of myth, whether in pained 'disillusionment' or in ecstatic endorsement of national conflict as cultural-racial 'hygiene' (for instance in the eugenic militarism of Karl Pearson, or the futurist claims of Marinetti for whom in 1915, 'War [was] the only cure for the world').

My point here is not to make the (absurd) claim that wars, least of all perhaps the First World War, should be seen as simply continuous with a 'pre-war' culture and society, but to suggest that a highly idealised juxtaposition of past and present may come to operate, imbuing war with a range of philosophical, evolutionary and psychological functions which eerily echo the war philosophy of the nineteenth century. These need to be analysed, not taken for granted. The First World War cannot be seen as either the final signified or the new signifier which emerges out of nothing – and yet so often it is.

Many different claims are made about 1914–18 as the prototype, model and threshold of modernity. According to Paul Fussell in his influential book, *The Great War and Modern Memory* (1975), the passing of the Military Service Act at the beginning of 1916 and the subsequent training of a conscript army represents an event 'which could be said to mark the beginning of the modern world'.[30] Haig, we learn, is the 'prototype' of subsequent images of the incompetence of military leaders:

> Indeed, one powerful legacy of Haig's performance is the conviction among the imaginative and intelligent today of the unredeemable defectiveness of all civil and military leaders. Haig could be said to have established the paradigm. His want of imagination and innocence of artistic culture have seemed to provide a model for Great Men ever since. (p. 12)

The First World War, Fussell powerfully argues, inaugurates a strange dehumanising vocabulary which has its late twentieth-century apotheosis in the euphemisms of military language: from 'combat fatigue' to Ministry or Department of 'Defence' (instead of War), from 'protective reaction strikes' to 'pacification centres' (in

30 Fussell, *Great War*, p. 11.

place of that earlier and increasingly unusable Boer War phrase, 'concentration camp').[31] The First World War, he contends, is a fundamental store of modern memory and subjectivity.[32] It opens up a whole bank of memory, a new nomenclature, even, as it were, a new unconscious of modernity: there are so many words, images and assumptions, now used without any overt or conscious reference to 1914–18, which have their etymological origins or particular connotations in the 'diction of war'. He cites amongst other words, 'crummy', 'lousy', 'rank and file' and 'No Man's Land' as examples.[33] Yet *The Oxford English Dictionary* indicates that none of these words actually originates in the First World War. They are all older. War language, as Fussell also acknowledges, drew on earlier models, images and metaphors of conflict and crisis. It is important to consider not only the First World War and subsequent modern memory, but the war as the locus of prior memories and forgetting.

Earlier descriptions and models of war contained many of the terms and elements of the characterisation of 1914–18. Kipling, for instance, was crucial in shaping the very image of the unromantic but patriotic, disenchanted but salt-of-the-earth imperial soldier in the 1890s. He focused on the sense of extreme divide between the world of military action and leisured contemplation, between war as activity and as vicarious enjoyment or mere speech.

The very designation 'Great War' was a common term in pre-war fiction as most obviously in *The Great War in England in 1897* by Le Queux or *The Great War of 189–. A Forecast*, written by Rear-Admiral Colomb and a group of other military and naval specialists.[34] The notion of the coming conflict as absolute moment of the modern and of historical differentiation was widely discussed. The 'surprise' of 1914 must be set against, or rather, linked to, the structure of earlier anticipation, the uncanny knowing/not knowing of the imminence of war.[35] The very notion of the 'Great War' as the 'war to end all wars' was a powerful concern in pre-war

31 See *ibid.*, p. 17.
32 As one review quoted on the back cover of the paperback edition of *The Great War and Modern Memory* enthusiastically declares, '[Fussell's book] is an examination of the war that put an end to the innocence of mankind and changed all human consciousness for ever'.
33 Fussell, *Great War*, p. 189.
34 For further discussions of these texts, see Clarke, *Voices Prophesying War* and Eby, *The Road to Armageddon*.
35 For an interesting discussion of this question of surprise and anticipation in France in 1914, see Becker, 'La Genèse de l'Union Sacrée'. For England, *cf.* Eby, *The Road to Armageddon*.

fiction and military speculation. Images and characterisations of the 'Great War' inevitably draw on earlier writing and perception.

In its apparent assumption of a generational 'collective mentality', Fussell's cultural study strangely echoes the model and inverts the argument of earlier social psychological investigations of the First World War. If we look at Caroline Playne's studies of collective mentality in the 1920s, *The Neuroses of the Nations* and *The Pre-War Mind in Britain*, for instance, we can see in much clearer and indeed cruder terms such a positing of the national spirit, psyche or memory. Playne's concern, however, was not with 1914 as beginning, but as end point in a nineteenth-century process of degeneration. Her method for understanding the culture of war depends upon a theory of the degenerate crowd and the hysterical mass mind which had crystallised in the 1890s, for instance in the work of Gustave Le Bon. Modern man, she argues, has been driven mad by the intolerable burden of technology, industry, democracy and urbanisation: 'His nerve ends receive sense stimu-lations showered on them, shot at them in continual torrents, whilst his grandfather's nervous system had to deal with sense-stimulations arriving in what, by comparison, may be likened to a slow-flowing stream.'[36] Without the traditional supports, including those of religion, 'The demon of neurasthenia' bursts forth in a modernity besieged by too many books, parties, ideas and tech-nological changes. The 'relapse' of the war and its accompanying shellshock catastrophe follow from a pre-war collective disturbance – 'neurasthenical derangements', 'social and national insanity', precipitated inevitably by modern life and its 'breathless conditions' (pp. 6, 7, 38). Industrialisation makes man ever more machine-like: 'Life becomes a mechanical push without human emotional variation to relieve the pressure' (p. 41).

In Playne's view France was plagued by irresponsible militarism combined with *ennui*, lethargy and degeneracy until, phoenix-like, it found its new regenerative spirit in the early twentieth century. Although French and German collective neurosis are seen as intertwined, it is in that latter deeply 'immature' nation that the crisis of war madness has been particularly acute, and the symptoms especially gross:

> Mass neurosis had so smitten, numbed and locked up the hearts and minds of the ruling personalities of the Central Empires that they believed themselves to be swept into the war *automatically*.

36   *The Neuroses of the Nations*, p. 5.

... Men renounced reason and will in order to become automata at the bidding of their own ghostly fears. ... The momentum was too strong for anyone, too strong and universal. The reins of governance were actually and, in very deed, dropped on the back of galloping mass folly. (pp. 455–64)[37]

The Prussians, with their war cult, nervous disorders, persecution mania and deification of power (all these mental disturbances exacerbated by the Slavic elements within Prussian blood) produced a pan-German intoxication: 'They decked out their hysterical rage with the appellation "*furor teutonicus*".'[38]

The only faint hope for the future that Playne can hold out lies in the fledgling discipline of 'preventive psychiatry', drawing on Le Bon, Trotter, MacDougall, Hobson and other recent pioneers of crowd science. Such a science might perhaps prevent 'the terrible complex' of war from developing in the human mind (p. 8). Playne's title, *The Pre-War Mind in Britain*, presupposes that there was indeed such a thing as a representative national 'mind'. Moreover it is assumed that this 'collective mentality' was the fundamental cause of war – 'mob minds' cause 'international upheavals'.[39] Modernity has supposedly led to this social and psychological disintegration. Individuals have become more and more like machines. The very speed of communications produces alienation. Machines are unnatural. To apply 'mechanical appliances to the means of locomotion' is unhealthy: 'Distant and lonely places became suddenly accessible and in towns people were flung in masses through the streets or through underground tunnels in engine-driven vehicles' (p. 30). All those well-meaning and antiquated conferences at the Hague failed to understand the crisis of mentality produced by modern life, this 'streak of perversion' running through the contemporary world. The Hague represented 'the old order of

37 Cf. 'In France, chauvinistic nationalism took the form of a fatalistic mysticism darkly robed in a pessimism which bordered on religious melancholia. In Germany a rude, barbaric exaltation of force and the State as a deified personality claimed many as victims of an untimely obsession' (*The Pre-War Mind in Britain*, p. 19).

38 *The Neuroses of the Nations*, p. 73.

39 *The Pre-War Mind in Britain*, p. 16. This is a view which has been importantly questioned in the recent historiography of the First World War. For a major study of popular opinion in France, which challenges the idea that public enthusiasm was the cause of entry into the war, see Becker, *1914*. Becker finds manifestations of enthusiasm quickly generated and mobilised after the event.

things', failing to grasp the phenomenon of the crowd, in which individual minds 'impregnate' each other, and a collective regression occurs. The nineteenth century seemingly deranged the collective body and mind, making war inevitable. In this interpretation the First World War is not the beginning of a modern consciousness, but the warped effect of a deranged pre-1914 psychopathology.

Paul Fussell certainly points out that 'the [First World] war was relying on inherited myth', not only 'generating new myth' which has become 'part of the fibre of our own lives'.[40] But his study in fact says relatively little about those past myths. We are offered some descriptions of the uncanny premonitions to be found in the years immediately preceding 1914 (Hardy, for instance, talking ominously of graves and death in 1913), but a rather static relationship between pre- and post-1914 remains entrenched in the book. Despite the talk of earlier languages on war, 1914 is still seen in a somewhat mystifying fashion as an absolute psychological transformation and awakening: 'One reason the Great War was more ironic than any other is that its beginning was more innocent' (p. 18). 'Never such innocence again' as Larkin puts it in a phrase, Fussell in a subheading (*ibid.*).

Fussell's concern, he tells us in the preface to his book, is the way the dynamics and iconography of the First World War have provided crucial political, rhetorical and artistic determinants in subsequent history. Irony, he goes on, became the structuring mode of war representation. Again and again Fussell stresses discontinuity with the world pre-1914. Hence he refers to the way the First World War 'reversed the Idea of Progress'. The modernism of the war lay in its challenge to the very notion of an unfolding history – or even a sequential history of campaigns. Fussell suggests that there were really no 'battles' (in the traditional sense) any more – they are simply the convenient markers we use to bring the unheard-of extensiveness of the carnage into line with previous war language. Hence the impressive-sounding demarcation of 'occasions' like Loos, Verdun, the Somme and Passchendaele. This is fallaciously to suggest, argues Fussell, that such 'events' parallel Blenheim and Waterloo.

Whilst *The Great War and Modern Memory* is undoubtedly a rich, moving and evocative study, it sustains too easily the image of pre-war society as an antique, quaint place with simple values and ideas awaiting their wartime decimation. Fussell again: 'the Great War took place in what was, compared with ours, a static world,

40   Fussell, *Great War*, p. ix.

where the values appeared stable and where the meanings of
abstractions seemed permanent and reliable. Everyone knew what
Glory was, and what Honor meant' (p. 21). Conversely, he implies,
nobody knew afterwards.[41] Both human experience and literature
in the pre-war world are cast as morally simple, childlike and naive;
subsequently they become sophisticated, inflected by a modernism
of which the world apparently knew nothing earlier:

> Indeed the literary scene is hard to imagine. There was no *Waste
> Land*, with its rats' alleys, dull canals, and dead men who have
> lost their bones: it would take four years of trench warfare
> to bring these to consciousness. There was no *Ulysses*, no
> *Mauberley*, no *Cantos*, no Kafka, no Proust, no Waugh, no
> Auden, no Huxley, no Cummings, no *Women in Love* or *Lady
> Chatterley's Lover*. There was no 'Valley of Ashes' in *The Great
> Gatsby*. One read Hardy and Kipling and Conrad and frequented
> worlds of traditional moral action delineated in traditional moral
> language. (p. 23)

Any sense that, say, a Hardy, a James, a Conrad, even a Kipling,
might powerfully and challengingly disturb what was in any case a
representation of stasis and fullness gets lost. 'One' read; and there
was only apparently 'one' canon, with 'one' aura to be felt. No
discussion here, for instance, of pre-war futurism with its rhetoric
of war, speed, violence, machinery.[42] Fussell's book moves strangely
between the rich evocation of myths, sensations, images or sounds
and, effectively, the reproduction of the one central myth he is
examining. The stilted tableau of an Edwardian leisured elite is
transformed into some generic pre-war consciousness or even universal social condition:

> Although some memories of the benign last summer before the
> war can be discounted as standard romantic retrospection turned
> ever rosier by egregious contrast with what followed, all agree
> that the prewar summer was the most idyllic for many years. It
> was warm and sunny, eminently pastoral. One lolled outside on a

41 In a recent PhD, Rosa Maria Bracco has made the important point that
    Fussell's canon of post-war modernist writers and his claims for their
    representativeness of a wider new 'ironic' culture ignores the extent to which
    popular fiction continued to stress, quite unironically, the meaningfulness,
    courage, companionship, fraternity, and worthwhile self-sacrifice of the war
    period. See Bracco, 'British middlebrow writers and the First World War,
    1919–39'.
42 By contrast see Hynes, *A War Imagined*.

folding canvas chaise, or swam, or walked in the countryside. One read outdoors, went on picnics, had tea served from a white wicker table under the trees. (pp. 23–4)

Note that telling pronoun 'one' and the swift abandonment of his own caveat, 'although', at the beginning of the quotation. Any sense of social division – any intimation of a harsh material dimension to the experience of the majority – is lost in the perpetuation of the image of a pre-war world about whose climate and pleasures supposedly 'all agreed'; the passive construction in the passage enables it to appear that no one 'serves' in this world, but everyone is 'served'.

As in so many evocations of 1914, a 'generation' is conjured up, all of whom are mysteriously of one mind: 'Out of the world of summer, 1914, marched a unique generation. It believed in Progress and Art and in no way doubted the benignity even of technology. The word *machine* was not yet invariably coupled with the word *gun*' (p. 24). Beyond 1914, according to Fussell, a whole series of juxtapositions and links emerge and become fixed in modern memory: 'Dawn has never recovered from what the Great War did to it' (p. 63). It is as though the war is never forgotten, erased or dissolved – our dawns, once pure, are now for ever and always contaminated by carnage. But surely it would be important to acknowledge a history of the First World War and modern forgetting; of the conflict as itself caught between the memory and forgetting of its own 'pre-history'; of the continuity rather than just the disturbance of symbols and memorials. National identities, it can be argued, are founded as much on forgetting as on remembering. Collective sacrifice, suffering and solidarity may all be important, but as Ernest Renan put it in his essay 'What is a nation?' (1882):

Forgetting, I would even go so far as to say historical error, is a crucial factor in the creation of a nation, which is why progress in historical studies often constitutes a danger for [the principle of] nationality. Indeed, historical enquiry brings to light deeds of violence which took place at the origin of all political formations, even of those whose consequences have been altogether beneficial. . . . the essence of a nation is that all individuals have many things in common, and also that they have forgotten many things.[43]

---

43 Renan, 'What is a nation?', p. 11. Renan's insistence on the original violence which constitutes states and which they need to forget, is reminiscent of one of Proudhon's central arguments in *La Guerre et la paix*; see Chapter 4.

Fussell's concern is always with the novelty of 1914, the sense of the war establishing new paradigms, as here for instance: 'Prolonged trench warfare, with its collective isolation, its "defensiveness," and its nervous obsession with what "the other side" is up to, establishes a model of modern political, social, artistic, and psychological polarization' (p. 76). Yet to insist that the First World War crystallised bellicose assumptions, political, economic, diplomatic and cultural processes with a longer history (one involving collective isolation, defensiveness, nervous obsessions and polarisation) is not in itself to state something new. The 'topos of inevitable war'[44] had developed powerfully before 1914 and has been extensively studied. Amidst the second final wave of the imperial scramble for overseas territory, international 'antagonisms' formed and re-formed. The Anglo-German rivalry was consolidated from the 1890s to the 1900s.[45]

The language of this antagonism, as I have argued, drew heavily on the terms of evolutionary theory. There is a danger, however, of singularising what was always plural: the social Darwinist evocation of war as the antidote to decadence, unfitness and degeneracy was never the exclusive voice of the pre-war period any more than it was the only possible variant of evolutionary discourse at any one time. Nevertheless, a web of new ideologies of war struggle, violence and political fatalism developed – a fatalism which, as Mommsen observes, 'finally became, in the face of mounting tension . . . a sort of independent factor in its own right'.[46] War had frequently been characterised as a cure for degeneration and lethargy. Even the German Chancellor Bethmann Hollweg, who in the view of many historians was reluctant to go to war, sympathised with the idea that 'the [German] people were in need

44 See Mommsen, 'The topos of inevitable war'.
45 Paul Kennedy locates 1901–2 as marking 'a watershed in British attitudes towards Germany which may be compared in both scope and importance with the alteration in the German view of England that had occurred in 1896–7. From the time of the Boer War onwards, it is possible to detect a growing British conviction that there existed a "German threat" or "German challenge" which had to be countered' (Kennedy, *The Rise of the Anglo-German Antagonism*, p. 251). Kennedy emphasises 1870 as a moment of transition, even though he also recognises fundamental continuities of Anglo-German diplomacy from the 1860s to the 1870s. Gladstone, for instance, deplored the war of 1870 (as of 1866), but made no practical intervention (*ibid.*, p. 22).
46 Mommsen, 'The topos of inevitable war', p. 24. On German literary fatalism before 1914, *cf.* Vondung, 'Visions de mort et de fin de monde'; on the culture of crisis in the years before 1914, see Barraclough, *From Agadir to Armageddon*.

of a war'.[47] Groups like the Pan-German League conducted a wide agitation campaign after 1912 against the 'physical and moral emasculation' that peace produced. War was cast as the antidote to both cultural castration and national sterility. That fatalism, Mommsen suggests, became almost a depressive acceptance in the German government in the last days of the peace: 'A fatalistic attitude which no longer believed in the possibility of being able to control the course of events, overshadowed the Reich leadership in the days and weeks following the murder at Sarajevo' (p. 32). Something important had to be experienced, accepted, resigned before the outbreak of war, something of a depressive (or manic) acquiescence in a particular logic, language and consciousness. Whatever its novelty, 1914–18 was also to see a complex recapitulation *and* reshaping of the codes, fictions, fantasies and philosophies of nineteenth-century culture, notably the perception of an absolute schism between the pre-industrial rural world and an atomised, rootless modernity, a conception of technological progress and scientific acceleration deemed inseparable from the exercise of destructive power.

47 Quoted in Mommsen, 'Topos', p. 26.

14 'The Revolt of the Machines'

I have focused on the representation of the deranged machinery of war, some of its continuities and new conceptualisations from the nineteenth to the twentieth centuries. First World War and post-First World War fictions of the machine entered into a dialogue with that earlier literature, rather than simply displacing and dissolving a conceptual world of war before 1914. I now want to extend my earlier discussion of the anarchic and destructive machine with two further examples, just beyond the First World War.

First, Karel Čapek's *R.U.R.* (Rossum's Universal Robots), produced shortly after the military conflict and in the shadows of widespread social and political convulsions, above all in Russia. The play turns on the question of whether a robot has a passion and a soul. Can it desire and defy? Is 'it' in fact adequately described as 'neuter' or only via some more complex constellation of the third person singular – she, he, it, id? Above all, can robots become deranged? The managers of the machines are confident at the beginning of the play that the robots can be controlled, that these metal contraptions are not caught up in the human drama of desire or sexual attraction, and moreover that they are ostensibly free of resentment. But soon it transpires that the machines can in fact catch the disease of love and, worse still, can be infected by a boundless envy and madness. They are far from inanimate, or at least their inanimate nature does not free them from emotion.

At first the problem had seemed to be the insensibility of the robots. Suffering has to be brought in for industrial reasons; pain nerves are introduced to stop them mindlessly destroying or damaging themselves. Partly in response to Helena's humanitarian desire to protect the 'rights' of the robots, Dr Gall is induced to give them powers of irritability, the capacity to feel. Hence they cease to be robots. They start to know pain, to sweat, to fear, to reproach, to lament their own sterility. They begin to smash things up, to gnash their teeth, foam at the mouth, becoming 'quite mad. Worse than an animal.'[1] Not to put too fine a point on it:

> Occasionally they seem to go off their heads. Something like epilepsy, you know. It's called Robot's cramp. They'll suddenly

1 Čapek, *R.U.R.*, p. 51.

sling down everything they're holding, stand still, gnash their teeth – and then they have to go into the stamping-mill. It's evidently some breakdown in the mechanism. (p. 46)

The tale within the play transports us from the era of *Frankenstein* to the modern world of technology, industry and war. The history of the robot, as given in the play, had begun when old Rossum, like Victor Frankenstein, dabbled with the secret of living matter, endeavouring to create life from a test-tube with ever greater hubris: 'He wanted to become a sort of scientific substitute for God. He was a fearful materialist' (p. 14). The old inventor works feverishly to engender 'artificial living matter', obsessed by this vainglorious and even mad desire: 'old Rossum was mad.... The old crank wanted to actually make people' (p. 13).

But Old Rossum is already out of date. Caught up in his pure experiments, his reverie of divine creativity, he remains blind to questions of industrial exploitation and profit. By contrast, Young Rossum is a good engineer; he abandons the idealist search for an artificial man and turns to something anatomically simpler – a 'working machine' with no emotions, a new cheap intelligent worker, technically perfect, faultless in memory, but utterly soulless. A new industry emerges, a vast factory complex for the construction of the robot. Livers, brains and bones are manufactured on site; spinning mills are set up for the production of nerves, veins and digestive tubes.

Čapek's *R.U.R.* is written and participates in the age of Taylorism, of cost-benefit analysis, mass production, a new technocratic quest for maximum 'efficiency'. The robots are carefully costed at three-quarters of a cent per hour. As production costs fall, the management, surpassing even the Taylorist faith in future human emancipation, triumphantly insists on the limitless progressive possibilities: 'Everybody will be free from worry and liberated from the degradation of labour' (p. 51). That unemployment will increase vastly is deemed beneficial – human beings will henceforth exist simply to perfect themselves.

But soon the robots are caught up in the human economy of war, turned into soldiers in their millions. Moreover, they begin to move out of control, to capsize their subordinated relationship to the human beings.[2] Even the robot librarian goes mad with curiosity

---

2 This theme endured in Čapek's work. In his later book *War with the Newts* (1936), extraordinary tailed sea creatures the size of ten-year-old children (the 'newts') are discovered, exploited and enslaved by human beings. They are eventually armed and pushed into wars on behalf of their masters, but they

and rebellious desires; he reads everything and discovers in the end that 'I don't want a master. I want to be master over others' (p. 90). As Helena puts it: 'It's as if something was falling on top of us, and couldn't be stopped' (p. 79). The robots declare man to be a parasite on the robots and a whole system of power is shown to unravel. This dialectic of the master and the slave proves intrinsically unstable.

From somewhere in the Balkans war re-emerges. The inhuman robot soldiers spare nobody. Millions are killed; the human population dwindles, no births are recorded. It is the firearm, rather than simply fire itself or the word, which has opened up in the robots the desire and the capacity to be masters:

> *Third Robot*: You gave us firearms. In all ways we were
> powerful. We had to become masters!
> *Radius*: Slaughter and domination are necessary if you would be
> human beings. Read history.

In an increasingly desperate bid to offset the danger of these 'universal' robots, the human survivors contemplate introducing difference into the world of the homogeneous machines. In place of uniform models:

> each of the factories will produce Robots of a different color, a different language. They'll be complete strangers to each other.
>
> They'll never be able to understand each other. Then we'll egg them on a little in the matter of misunderstanding and the result will be that for ages to come every Robot will hate every other Robot of a different factory mark. (p. 113)

The war waged by the robots against human society can only be brought to an end by kindling discord within the technology, which finally ignites into a great internal battle of the machines. Their mechanical empire unravels in the bitterness of internecine struggle, the conflict of a thousand competing robot nationalisms.

Such a narrative could be loosely linked back to earlier myths and stories, like Faust or Frankenstein; or it could be related to a wider nineteenth-century critique of the alienating effects of the machine. But it also joined a contemporary debate about the nature of the relationship between human breakdown and modern war conditions which, I have suggested, crystallised in the half-century after 1870.

---

subsequently become uncontrollable and counterattack – indeed they literally undermine the established order by drilling through the earth, causing earthquakes and landslides.

This problematic issue of control in the machine age was also central in the (unrealised) film script which emerged from the collaboration of Romain Rolland and Frans Masereel in 1921, entitled *La Révolte des machines*. The story begins once again with human lordship over technology. We are presented with a huge ensemble of machines and, entering on a conveyor belt, a cortège of human dignitaries with their somewhat grotesquely regal manners.[3] Their president makes a speech to the gathered assembly, a eulogy to science, civilisation and enlightenment. Modernity, he enthuses, has soared forth from old darkness and obscurantism, eliminating the illogical and the inefficient; agriculture today has become a colossal industry. Enormous areas are maintained with startling rapidity by machines and regulated by a single man seated at a strategic observation point who is able to spend most of the time reading his newspaper. Progress is a zigzagging stream, at times apparently blocked, but in reality always moving forward to an apotheosis where the godlike human being will control everything from an armchair, at the flick of an electric switch. It is a hymn to progress, human genius and lordship over nature.

At this point in the script, a dance of the machines begins. With the push of a button by a character called Martin Pilon, the engineer and 'master' of the machines, the engines start to move 'in a Prussian fashion'. Pilon appears as part wizard, part army commander, putting the troops through their paces to the applause of the crowd. Yet he is an unbalanced figure, his own internal machinery out of control, his body fired up with 'subconscious electricity'. He has charge of a variety of machines. Some are built for power, others for their spider-like dexterity or psychological penetration. One machine is even able to read the thoughts of surrounding people. It has the form of an eye at the end of an extensible elephant-like trunk, which it is able to clamp on to the skull of its patient/victim. The subject's thoughts are then projected on to a screen.[4] When the machine is surreptitiously locked on to Pilon, he is profoundly annoyed and, having failed to control the device, becomes enslaved by his own submerged angry impulses: 'In his irritation he loses control of himself and his subconscious begins to come into play. It is the beginning of the revolt of the machines.'

The story dramatises the collapse of human authority and self-mastery. In part this is a simple transfer of power: masters and slaves change places; human beings are turned into objects, machines of the machines. But the reversal produces a more general derange-

3 Rolland and Masereel, *La Révolte des machines*, p. 35.
4 See *ibid.*, p. 55.

ment too, a disintegration of all control and rational agency amidst the ever more outrageous antics of the machines. First the moving pavement begins to shake about in order to throw off its passengers; machines play pranks, pinch noses, blow smoke into people's faces or pull their coats up their backs. Finally they lift the president helplessly into the air, despite the entreaties of Pilon, who is eventually held responsible and, to his fury, carted off to gaol. The machines throw off their bonds and escape into the night, panicking passers-by. The 'movement of the great monsters' is under way.

The machines release Pilon from his incarceration and he is seen endeavouring vainly to reimpose his authority. But they have now become unequivocally destructive forces that turn their anguished master into their impotent slave. Destruction gathers pace; the town crumbles and collapses 'like a house of cards' (p. 80). The machines take on animal qualities — toying with their former masters like sniffing dogs only to rush away again with terrifying and crazy determination. They destroy the landscape — the fields, crops, trees around them — to the helpless horror of human observers. The people send tanks against the machines, but the vehicles change sides and take the tank crews prisoner. Under this attack — 'on all sides, machines, machines' — civilisation begins to buckle and to cave in. Amidst the mêlée of planes, tanks and trains, a kind of psychosis is in the air ('a frenetic whirlwind', a 'delirium', p. 95) which is not dissolved by the eventual rather anodyne denouement in which the people escape the attack by sowing discord amongst the machines and by dismantling some of the motors. Humanity returns miraculously to 'nature', a machineless world of 'rustic joy', of milk, earth and honey.

Just before the outbreak of the First World War, John Broadus Watson, a professor at Johns Hopkins University in Baltimore, had declared: 'The time seems to have come when psychology must discard all reference to consciousness.'[5] Watson's methodology for the study of the human being imagined itself 'a purely objective experimental branch of natural science' and aimed at 'the prediction and control of behaviour' (p. 158); it could not tolerate any 'intangibles and unapproachables'.[6] 'Behaviourism' was to do away with introspection and the unquantifiable. Not by chance was this project to emerge in the age of Taylorism and mass production, and in a period where history itself was proving so disturbingly untamable, its outcome so ineffable and complex; and at a moment

5 Watson, 'Psychology as the behaviourist views it', p. 163.
6 Watson, Behaviourism, p. 6.

when the propagandistic drive to establish sharp and arbitrary boundaries between legitimate and illegitimate national identities and knowledges was gaining so powerful a grip. So many branches of learning would seek to purge themselves of the Germanic, not least behaviourism, which aimed to exclude the introspective techniques so deeply associated with German psychology.[7]

In his famous critique of behaviourism, the system built by B.F. Skinner on Watson's foundations, Arthur Koestler argues that what is left after this reductivist purge is little but rats. Behaviourism, he suggests, is the heir to nineteenth-century science at its most mechanistic and atomistic. Yet the human being is irreducible to the world of reflex responses and behavioural conditioning; Koestler insists on the ghost in the machine, speculating on 'the streak of insanity which runs through the history of our species' and pondering whether human evolution perhaps 'contains some built-in error or deficiency which predisposes [man] towards self-destruction' (p. xi). The human being, he insists, cannot be understood as a mere automaton possessed of slot-machine reflexes. In short, a true 'science of life' cannot be mechanistic (p. xiii).

But what the fictions of Čapek and Rolland suggest is not just a ghost in the human 'machine', but an irreducible realm of desire, aggression and madness in the technology fashioned by the human being. The aim of these examples has been to bring into sharper relief some of the key images of the war machine as they emerge and develop between the nineteenth and early twentieth centuries, questions of technological control, political power, human determination; imputations of a will, even an unconscious, in the robot, and, conversely, of mechanical, behaviouristic flaws in the human being, a set of representations of friction, destruction, chance, desire and identity which resonate far beyond Watson's behaviourist agenda.

---

7   See Koestler, *The Ghost in the Machine*, p. 17.

15 'Why War?'

After that fleeting excursus into Watsonian behaviourism, and before turning to Freud on war, a word on his one-time disciple and subsequent apostate Wilhelm Reich. 1933 – the year of publication of the Freud–Einstein dialogue *Why War?*[1] – was also to see the appearance of *Reich's Mass Psychology of Fascism* which, in moving between Marx and Freud, insists that world war cannot be explained away economically, nor can it exclusively be understood psychologically. Reich instead appeals to history, by which he intends the notion of millennia of perversion, involving the substitution of money for love, of power fantasies and racism for freedom.

In Reich's study the question of questions is 'What produced the *mass psychological soil* on which an imperialistic ideology could grow and could be put into practice, in strict contradiction to the peace-loving mentality of a German population uninterested in foreign policy?'[2] Fascism, he insists, is not an imposition by force, but the politically organised expression of the average human character structure in authoritarian society, with its machine civilisation and its mechanistic view of life.

As in so many of the stories and histories already discussed, machinery has a central place in this account. Reich argues that the machine has been idealised and fetishised, whilst simultaneously the 'little man' becomes a kind of automaton, biologically crippled and capable of 'machine murder'. Economically oppressed, sexually cramped and repressed, the masses affirm their own subjugation under Fascism.[3] Women become simply 'child-bearing machines'. Again, it is the Prussian who best epitomises this deformation of the human into the mechanical: 'The Prussian military display shows all the characteristics of the mystical machine-man' (p. 295). But more generally too, history has increasingly witnessed a primitive cult of the machine. The locomotive is a kind of totem imbued with 'eyes to see and legs to run, a mouth to eat the coal and excretory

---

1 See the next section, pp. 214 ff.
2 Reich, *The Mass Psychology of Fascism*, p. 17.
3 For a compelling if problematic extension of this perspective, see Theweleit, *Male Fantasies*.

apparatus for the slags, and [a] mechanism for the production of sounds' (p. 287). Humanity has had the perennial dream that the machine would make life easier; but in reality the machine has become a deadly enemy, above all because it enables this insane projection of human desire. The human body is shaped by the world of machinery, and human knowledge limps along in the same direction: mind is reduced by an impoverished modern psychology to a matter of mere secretions in the brain. This degradation of both existence and knowledge culminates in the distorted view of war as itself merely a function of the irresponsible machine:

> The tragic split between biological and technical understanding, between that which is alive and that which is machine-like in man, is unequivocally expressed in the following: No mass individual in the world wanted the war. But all are its victims, as of a mechanical monster. *Yet this monster is the biologically rigid human himself.* (p. 289 n. 1)

To summarise Reich's thesis: humanity is biologically sick; politics is the irrational social expression of this illness; the masses are the decisive factor and Fascism the expression of their alienated desire. Nazi gangsters may constitute an outrageous swindle, but one unfortunately affirmed by the adoring populace. The human structure is determined by the strife between the longing for freedom and the terror of freedom, which over the last 4–6,000 years has been, to say the least, a rather unequal conflict in favour of the latter (see pp. 274–5).[4]

For Reich, as for Schumpeter, economic motives are quite insufficient to understand what is at stake in imperialism and war. The middle classes, for instance, do not behave politically in a way that reflects their interests, or at least their interests as narrowly conceived: 'Always looking above himself, the middle class indi-

---

4   *Cf.* Erich Fromm's development of this thesis in *The Fear of Freedom* (1942). Fromm speaks of 'automaton conformity' and an unprecedented alienation in the modern world. Nazism and Fascism need to be understood psychologically (we must locate the psychic compensations they offer). It is quite insufficient, he insists, just to explain them away economically. In *The Anatomy of Human Destructiveness* (1973), he again explores the peculiar alienating and fragmenting propensities of a modern machine age, contrasting the antiseptic gleaming face of weaponry with the destruction missiles unleash. Indeed in modern war, the human agents of the destruction (above all pilots) are utterly separated from the victims. 'Once this process has been fully established there is no limit to destructiveness because nobody *destroys*; one only serves the machine for programmed – hence, apparently rational – purposes' (p. 348). Modern human beings, indeed modern societies at large, Fromm continues, are prone, to a quite unprecedented degree, to *schizophrenic* qualities (see pp. 351–7).

vidual develops a *divergence between his economic position and his ideology*' (p. 39). Militarism itself must be understood as fundamentally libidinous; Fascism as fundamentally phobic: 'The irrational fear of syphilis is the most potent source of National Socialist political Weltanschauung and anti-semitism' (see pp. 26, 69). Nor can 'nature' be used as the ultimate explanation since 'There are in the animal world, no wars between members of the same species' (p. 271). For Reich, the war drive does not originate in some archaic force or instinct, nor, as in Freud, does it lead to a pessimistic recognition of the intractability of some of the impulses underlying aggression. For Freud internal factors cannot be abolished by future decree or utopian plan, which is not to say that society cannot perhaps be better organised to anticipate war dangers. For Freud there are certain irresolvable internal conflicts within the subject, and between the subject and society, whilst in Reich's account there has been a long-drawn-out collective perversion and socially induced derangement which had a beginning *and could have an end*: 'As a result of thousands of years of social and educational warping, the masses of the people have become biologically rigid and incapable of freedom. They are no longer capable of organizing a peaceful living together' (p. 271). For Reich, this 'warping' is epitomised in modern society by the emotional investment in the machine, and in the behaviouristic reduction of the psyche to the mechanical ethos:

> What is called civilized man is in fact angular, machine-like, without spontaneity; it has developed into an automaton and a 'brain machine'. Man not only believes that he functions like a machine, *he does in fact function like a machine*. (p. 293)

Quite unlike Freud, Reich imagines a potentially definitive social freedom, achieved through 'work democracy'. This could constitute a liberation from the phobias, fantasies and identifications he catalogues. Reich's political assumptions did not adhere to those of the Soviet route. He did not assume the mantle of Pavlovianism, but rather a libertarian conflation of Marxian revolution, ecstatic sexuality and what Elizabeth Roudinesco calls 'delirious biologism'.[5] Thus he exhorts the reader 'to give up the irrational denial of his own nature' (p. 290). To re-find that nature, it seems to Reich (not to Freud), might be to re-find a deep and perpetual peace.[6] But then

---

5 Roudinesco, *Jacques Lacan and Co.*, p. 45.
6 *Cf.* Freud's comment on Reich in 1928: 'We have here a Dr Reich, a decent but impetuous mounter of battle horses who now venerates genital orgasms as the antidote for all neurosis' (quoted in *ibid.*, p. 44).

Freud was only half a Bolshevik, sharing that sect's sense of current gloom and prospective crisis, but not, he insisted, its future consolations of collective peace, prosperity and happiness.[7]

FREUD–EINSTEIN

In his correspondence with Freud in 1932 (subsequently to be entitled *Why War?*),[8] Albert Einstein freely acknowledged that he was troubled by uncertainties about the causes and deep meaning of military conflict. He turned towards Freud, by his very gesture apparently repudiating any confidence that natural science alone could fully explain war or even claim any special status in resolving the problems of human conflict. Freud later touched on the ambiguous nature of the authority of his interlocutor in this debate by suggesting that Einstein 'had raised the question not as a natural scientist and physicist but as a philanthropist' (p. 203). The inquiry culminated in a question to which Freud subsequently made a somewhat half-hearted reply: 'Is it possible to control man's mental evolution so as to make him proof against the psychoses of hate and destructiveness?' (p. 201)

The discussion had arisen from an initiative in 1931 when the International Institute of Intellectual Cooperation was instructed (by the Permanent Committee for Literature and the Arts of the League of Nations) to arrange for exchanges of letters between representative intellectuals on subjects that might serve the common interests of the League and of intellectual life. As a result, Einstein was invited to address a public letter to a correspondent of his choice, the results subsequently to appear in print. Einstein chose to put his question to the pioneer of psychoanalysis. The exchange of letters between the physicist, writing from Caputh near Potsdam, and the psychoanalyst in Vienna, took place between August and September 1932, and was published the following year by the Institute of Intellectual Cooperation, in German, French and English simultaneously. Its circulation was forbidden in Germany.[9]

---

7 See Jones, *Sigmund Freud: Life and Work*, vol. III, p. 17.
8 See editorial introduction to Freud, *Why War?*
9 Strikingly, at about this time, when pressed by the father of one of his patients, who was a friend of Mussolini's, to send an example of his work with a dedication to the *Duce*, Freud chose 'Why War?' Freud's dedication offered 'the devoted greeting of an old man, who recognises the cultural hero in the ruler', an allusion according to the Italian analyst Edoardo Weiss, who was responsible for bringing the patient, to Mussolini's major archaeological excavations in which Freud was keenly interested. See Gay, *Freud*, p. 448n.

Einstein raised doubts about the capacity of natural science to understand, let alone solve, conflict and aggression in society; as though, in the wake of the First World War and in the gathering gloom of Nazism, it was necessary to address more humbly than ever the question of the limits of a positive science of the social. Freud pushed this vexed issue still further when, in touching on the relation of war to the death drive, he admitted that his conjectures might appear close to mythology, but then in fact 'does not every science come in the end to a kind of mythology?' (p. 211).

Mythology or not, each of the correspondents in this self-conscious discourse of great minds sought to exclude himself from complicity in the 'war fever' under discussion. Freud argued some-what against the grain of his own work and many of his own pronouncements in the published letter itself, that 'we pacifists have a *constitutional* intolerance of war' (p. 215). Ambivalence and contradiction, in short, are central not only to Freud's theory of war, but also to his own personal response to it. Einstein also claimed to be 'immune from nationalist bias' (p. 199), whilst simultaneously rejecting the old idea that a scientific or rational elite was better able than the majority to resist the psychological lure of war. He insists that the susceptibilities of the educated individual and the hypnotised masses cannot easily be differentiated. Indeed, if anything, the intellectual and the educated were more vulnerable than the uncultured population to collective manipulation:

Experience proves that it is rather the so-called 'intelligentsia' that is most apt to yield to these disastrous collective suggestions, since the intellectual has no direct contact with life in the raw, but encounters it in its easiest synthetic form – upon the printed page. (p. 201)[10]

10 Even before the First World War, and ensuring bitter reflections on the 'treason of the intellectuals', the idea of the peculiar war susceptibility of the educated had been explored in the language of a self-critical liberalism. *Cf.* Hobson's *Psychology of Jingoism* (1901): 'Now the most astonishing phenomenon of this war-fever is the credulity displayed by the educated classes' (p. 21). After 1914–18, of course, that perception grew much stronger. To take another example from the 1930s, note how Virginia Woolf insists that certain kinds of education offer no protection, no civilisation against force, possessiveness, war: 'Need we collect more facts from history and biography to prove our statement that all attempts to influence the young against war through the education they receive at the universities must be abandoned? For do they not prove that education, the finest education in the world, does not teach people to hate force, but to use it? Do they not prove that education, far from teaching the educated generosity and magnanimity, makes them on the contrary so anxious to keep their possessions, that "grandeur and power" of which the poet speaks, in their own hands, that they

Print or writing are contrasted with the 'raw' life, which is less likely to mislead; intellectuals are more liable to be transported to strange mental places than others, non-readers presumably, who are deemed more capable of keeping their feet firmly on the ground. The correspondents were evidently at odds on this issue of elite and mass susceptibility. Could 'leaders' and 'led' be distinguished? Einstein argues that the reactions of an elite could not be separated from those of the masses in the face of war fever; Freud observes that 'One instance of the innate and ineradicable inequality of men is their tendency to fall into the two classes of leaders and followers' (p. 212).

The cure for war, Einstein reflects, in one way appears obvious: the establishment of an international legislative body whose decision will be invoked and found binding in every dispute. Nations will voluntarily sacrifice their own 'might' in pursuit of this collective 'right'. Unfortunately the problems are enormous, not least human psychology itself: 'The craving for power which characterises the governing class in every nation is hostile to any limitation of the national sovereignty' (p. 200). 'Political-power-hunger', 'personal interests' and the complex relationship between the elite and the masses form part of the problem. Einstein echoes a cosmopolitan dream of peace, evoking the idea of a binding international court of arbitration, but only to turn back to Freud with a recognition of the intractable problem of the human irrational. War, like love, cannot be neatly channelled by Leagues, pacts or agreements. Einstein is concerned with propaganda, but also seems aware of the inadequacy of any model which focuses simply on external manipulation. Certainly, the elite, habitually in control of the church, the press and education, is able 'to organise and sway the emotions of the masses, and make its tool of them' (p. 201). But how does propaganda work? 'How is it possible for this small clique to bend the will of its majority, who stand to lose and suffer by a state of war, to the service of their ambitions?' (p. 200). Beyond the question of political skullduggery and rhetorical distortion, there remains the still more difficult and ambiguous matter of the 'lust for hatred and destruction' itself. Whilst this can normally be contained 'in a latent state', 'it is a comparatively easy task to call it into play and raise it to the power of a collective psychosis' (p. 201).

Freud, it should be noted, confessed to finding the exchange rather tedious and sterile. There is indeed a certain weary tone, as though perhaps Freud felt the terms of the debate were a little

will use not force but much subtler methods than force when they are asked to share them? And are not force and possessiveness very closely connected with war?' (*Three Guineas*, p. 35).

anachronistic – two great intellectuals courteously exchanging views by correspondence on the general truths and universal illusions of war. There is a certain air of abstraction, which seemed to discomfort Freud even as he participated in it: at stake not Fascism, not the First World War, not just an impending war, but, precisely, 'Why War?' Where Reich at the same moment would address the mass psychology of Fascism before turning it back into some more widespread and long-lived phenomenon, Freud pitches his letter from the beginning at a formidably generic level of debate – the level of a speculative anthropology, history and ontology, which, whatever its fascination, its resonances, its value, its pertinent contradictions, is very different from the delicacy of interpretation so often insisted upon and demonstrated in his clinical and dream interpretations. He expounded his view that 'right' is anthropologically and historically secondary to 'might'. 'It is a general principle, then, that conflicts of interest between men are settled by the use of violence. This is true of the whole animal kingdom from which men have no business to exclude themselves' (p. 204). By contrast, remember Reich's insistence that animals know nothing of war against members of their own species; animal violence and human war must be firmly distinguished.[11]

In Freud's model, the violence of the strongest individual is met 'in the course of evolution', by the collective strength of the weak. Through the power of their union, their own demands became law: 'right is the might of the community' (p. 205). Society thus is founded on violence and conflict; it involves transposing power from the most powerful individual to the group. The collective violence of the weak has to be stabilised as a new legal order of 'right'. This stabilisation and perpetuation of the defensive community, of course never totally achieved, aims to prevent a full regression to individual violence, through, as it were, monopolisation or collective ownership of the means of production of violence:

The community must be maintained permanently, must be organised, must draw up regulations to anticipate the risk of rebellion and must institute authorities to see that those regulations – the

11 One might also record here the psychoanalyst Roger Money-Kyrle's observations in his book *Superstition and Society* (1939): 'There is, too, a vast gap between human war and animal aggression. Animals do not prey upon their own species. They squabble over food, the males fight over the females; but if the defeated rival runs away, he is not pursued and killed.... The essence of human war, however, is that each group fights together, under a leader. The conscious motives too are different and less rational. Of these the most fundamental is probably revenge – a motive apparently unknown to animals, except in just-so stories' (p. 105).

laws – are respected and to superintend the execution of legal acts of violence. (*Ibid.*)

Ideally, violence is continually transferred to larger social units – although in practice its distribution remains uneven and unjust. This supposed evolution of violence from individual force towards a kind of peaceful democracy of might is further undermined by social, familial and gender inequalities, and it is destabilised by the dynamics of mastery and slavery that wars bring in their wake:

> a state of rest of this kind is only theoretically conceivable. In actuality the position is complicated by the fact that from its very beginning the community comprises elements of unequal strength – men and women, parents and children – and soon, as a result of war and conquest, it also comes to include victors and vanquished, who turn into masters and slaves. The justice of the community then becomes an expression of the unequal degrees of power obtaining within it; the laws are made by and for the ruling members and find little room for the rights of those in subjection. (p. 206)

All of which led Freud to an uncomfortable admission:

> Paradoxical as it may sound, it must be admitted that war might be a far from inappropriate means of establishing the eagerly desired reign of 'everlasting' peace, since it is in a position to create the larger units within which a powerful central government makes further wars impossible. (p. 207)

Freud developed here a point he had made in his earlier essay 'Thoughts for the Times on War and Death' (1915) – that 'piece of topical chit-chat about war and death to keep the self-sacrificing publisher happy' as he dismissively wrote to Karl Abraham.[12] 'Thoughts for the Times' had addressed this question of monopolisation rather more bleakly than in the upbeat conclusion to his letter to Einstein.[13]

---

12 Letter of 4 March, 1915, in Freud and Abraham, *A Psychoanalytic Dialogue*, p. 213.

13 The letter to Einstein of 1932 declared that 'it may not be Utopian to hope that these two factors, the cultural attitude and the justified dread of the consequences of a future war, may result within a measurable time in putting an end to the waging of war' (*Why War?*, p. 215). For a further First World War 'chat' in similarly pessimistic terms to those of 'Thoughts for the Times', see Freud's talk to the Jewish Cultural Association, a section of the Lodge of B'Nai B'Rith. This paper, entitled 'Wir und der Tod' ('We and Death'), has only recently been published in *Die Zeit*; see the Bibliography for details.

In that earlier work from the period of the First World War, Freud noted the irony that scientific and technological advances produced ever more destructive capability. War today is 'more bloody and more destructive than any war of other days, because of the enormously increased perfection of weapons of attack and defence'. Moreover, that 'advance' has not been matched by a civilising of war itself, which remains 'at least as cruel, as implacable as any that preceded it' (p. 278). In 1915 the state's monopolisation of the means of war provoked a sense of horror rather than, as for certain late nineteenth-century theorists, of reassurance:

> The individual citizen can with horror convince himself in this war of what would occasionally cross his mind in peace-time – that the state has forbidden to the individual the practice of wrong-doing not because it wishes to abolish it, but because it desires to monopolize it, like salt and tobacco. (p. 279)

Cultural evolution and Eros, Freud would suggest later in *Why War?*, is responsible for man's partial revulsion against war. On the other hand sexuality is itself invested to a greater or lesser extent with aggression, with the destructive force of sadism and masochism. As Freud shows in *Civilization and its Discontents*, there is an enduring conflict between Eros and the death drive, yet these terms are adulterated with one another from the start.[14] Whilst 'There is no use trying to get rid of men's aggressive inclinations',[15] nevertheless Eros should be encouraged; anything that fosters the growth of emotional ties militates against war, Freud declares, although again the opposite formulation is also sometimes offered: emotional ties militate for war.

The trouble is that, according to Freud, there exists a continuing tension between culture and human survival. Whilst culture helps

14 'Starting from speculations on the beginning of life and from biological parallels, I drew the conclusion that, besides the instinct to preserve living substance and to join it into ever larger units, there must exist another, contrary instinct seeking to dissolve those units and to bring them back to their primaeval, inorganic state. That is to say, as well as Eros there was an instinct of death...a portion of the [death] instinct is diverted toward the external world and comes to light as an instinct of aggressiveness and destructiveness. In this way the instinct itself could be pressed into the service of Eros, in that the organism was destroying some other thing, whether animate or inanimate, instead of destroying its own self' (*Civilization and its Discontents*, pp. 118–19).

15 *Why War?*, p. 211. Freud noted the illusory claim of the Soviet Union to achieve this, even though, in his view, the society was partly kept together by hatred of those beyond its frontiers.

build up bonds inimical to war, it also disarms the individual's desire for aggression *from within*. The superego is like civilisation's fifth column, it sometimes appears; or as Freud puts it in *Civilization and its Discontents*, again drawing on a military metaphor, it is 'like a garrison in a conquered city'.[16] Culture also makes more likely the 'extinction of the human race' (p. 214) by progressively disjoining the human being from the species function of reproduction. It runs against instinct, producing an ever deepening crisis of renunciation and repression, a kind of irreducible historical and psychological friction. The 'process of evolution of culture', which accounts for 'the best of what we have become', is also responsible for 'a good part of what we suffer from'. This process 'may perhaps be leading to the extinction of the human race, for in more than one way it impairs the sexual function; uncultivated races and backward strata of the population are already multiplying more rapidly than highly cultivated ones' (p. 214).

By the late 1930s Freud's writing had come to engage still more insistently with the question of intractability. He suggests that there may be insuperable difficulties to 'cure' whether in relation to anti-Semitism or to the 'progress' of the individual patient. Thus *Analysis Terminable or Interminable* (1937) sombrely but stoically stresses the recalcitrance of the patient, of instinct, of the primitive: it accepts the inevitability of at least a partial therapeutic failure, or rather the redefinition of the objectives of the treatment. Freud spelt out, not for the first time, the fact that knowledge is not cure and that resistance is formidably and often necessarily persistent; analysis is potentially interminable, inevitably incomplete. Consider also the bleak conclusion of *Moses and Monotheism*, written between 1934 and 1938, discovering an anti-Semitism without end, unless, impossibly, remaining Jews renounced their religion and became Christian:

> It is worth noticing how the new religion dealt with the ancient ambivalence in the relation to the father. Its main content was, it is true, reconciliation with God the Father, atonement for the crime committed against him; but the other side of the emotional relation showed itself in the fact that the son, who had taken the atonement on himself, became a god himself beside the father and, actually, in place of the father. Christianity, having arisen out of a father-religion, became a son-religion. It has not escaped the fate of having to get rid of the father.... A special enquiry

16   *Civilization and its Discontents*, p. 124.

would be called for to discover why it has been impossible for the Jews to join in this forward step which was implied, in spite of all its distortions, by the admission of having murdered God. In a certain sense they have in that way taken a tragic load of guilt on themselves; they have been made to pay heavy penance for it.[17]

Whatever their relation to the internal history of psychoanalytic thought, these writings, I suggest, are indeed signs of the times, part of a wider set of cultural analyses. The consolidation of the power of Fascism and Nazism together with the catastrophe of the Spanish Civil War signified for so many writers in this period the closing of a window of hope (wherein the First World War would have represented a definitive moral lesson about the universal futility of war – as in the vastly popular and influential *All Quiet on the Western Front*, 1929). The glimmer of optimism was still just there perhaps in the very fact of Einstein's turn towards Freud and even at points in Freud's reply. Psychoanalysis after all knows something about war and rage in the subject, and also provides a special place in which they may be both faced and contained without the reflex demand of 'civilisation' that they be abolished by decree or educated away.

Diverse cultural commentators in this period take it upon themselves to reject pacifist 'naivety' not only at the level of practical politics, but anthropologically and universally; they argue against the strong appeal of appeasement and quiescence, perceiving an historical situation of receding peaceful aspirations, irrevocable tendencies to war, aggressive instinct, nationalist fanaticism and the triumph of tyrannical organisations over both the individual and the collective will. So many commentators in the 1930s felt themselves to be confronting an endemic economic crisis without apparent end; a dissolution of the flickering optimism of previous years, the forlorn Locarno 'non-aggression pact' of 1925, for instance, between France, Germany and Belgium, guaranteed by the supposedly neutral powers, Great Britain and Italy. Thus we encounter voices aiming to generalise once more the meaning of the First World War – understood as beginning rather than end of a crisis of modernity. They express the drama of a seemingly inexorable Fascist rise. Part of the function of war, as of Fascism, in psychoanalytic terms, is to abolish an internally ambivalent relationship, to locate all the hated components elsewhere and destroy them – the Jews, the enemy nation. It is necessary to fight

17 *Moses and Monotheism*, p. 136.

wars, it is argued, in order to find and only secondarily to destroy the enemy.

I have already introduced various war commentaries to set alongside Freud's, but let us take, as one further brief example of the wider and intensifying contemporary pessimism to which I allude here, the celebrated French historian, Elie Halévy. In 'The world crisis of 1914–18', his Rhodes Memorial Lecture at Oxford University in 1929, he speaks in ominous terms of the problem of military conflict, or as he put it, ' "strife", not "war", because the world crisis of 1914–1918 was not only a war – the war of 1914 – but a revolution – the revolution of 1917.'[18] He asks his audience to face their own English implication in a terrifyingly stubborn problem of nationality and hate, which, while nothing new, has become ever more dangerous and important in the present:

> National fanaticism is something far more formidable than class fanaticism. England has eliminated the one, but not the other. She may have been, during two centuries, a nation without a revolution; it can hardly be said that she has not been a warlike nation. Even during the last quarter of a century, when mankind has seemed more anxious than ever before to find some way out of war through arbitration and compromise, has any government, that of England included, subscribed to any peace pact, even to the Covenant of the League of Nations, without making, explicitly or implicitly, some reservation? . . . But the fact remains, that man is not wholly made up of common sense and self-interest; such is his nature that he does not think life worth living if there is not something for which he is ready to lose his life. . . . So long as we have not evolved a fanaticism of humanity, strong enough to counterbalance or absorb the fanaticisms of nationality, let us not visit our sins upon our statesmen. Let us rather find reasons for excusing them, if they occasionally feel compelled to submit to the pressure of our disinterested and fanatical emotions. (p. 190)

In his communication, 'The era of tyrannies', presented to the Société Française de Philosophie in 1936, Halévy insisted on the use of the Greek word 'tyranny' in preference to the Latin 'dictatorship', because in his view the former rightly conveys an habitual state of affairs, rather than a provisional regime. In short, 'tyranny' refers to a normal process in world history since 1914 – reflected in Fascism as well as Bolshevism (indeed in the Fascism of Bolshevism). The twentieth century has witnessed the vast expansion of state

18 Halévy, *The Era of Tyrannies*, p. 162.

control of the economy (including the co-opting of the working class to the interests of the nation) together with an unprecedented state interference in thought itself (both of a proscriptive and an inciting 'positive' nature). The war erupted out of a gathering earlier crisis. We must not be 'nostalgic' about the pre-war period, he warns, or imagine it as somehow idyllically peaceful. Modernity has witnessed the progressive destruction of 'humane' barriers to violent change. Future war now seems virtually inevitable and 'If it comes, the situation of the democracies will be tragic. If they want to wage war effectively, can they remain parliamentary and liberal democracies? . . . When the war begins again, it will consolidate the "tyrannical" idea in Europe' (p. 213).

The Freud–Einstein exchange can be read as one particular, and particularly forlorn, summation of a longer-running debate or dialogue on war; but it also clearly captures, more narrowly, a specific and dwindling moment of political hope. I introduce this dialogue here both to draw attention to the moving resonances of its content and to remark on certain terms of discussion it excludes or takes for granted. It must be admitted that Freud's language rejoins at moments a long and highly contentious social Darwinist tradition,[19] which assumes the superiority and conversely the degeneracy of certain sections of the species. Freud uses too easily terms like 'backward strata of the population' or 'uncultivated races'. On the other hand, the letter touches in far less reductive ways on the function of war in binding society and the enduring social-psychological tension that exists between different drives and pressures. Thus for Freud the recalcitrant force of instinctual aggression continues to hurl itself against the barriers of 'civilisation'. Freud addresses the disparity between the sophistication of weaponry and the primitiveness of the actual conduct of war. He also raises the question of how far an aboriginal violence of the individual and group are grafted on to, even as they are apparently displaced by, the laws and codified wars of the community or the nation.

Freud's analyses of war alert us to the displaced primitive violence operating even in the most sophisticated war. In war, the state seeks to monopolise violence, to husband and marshal it. But other factors are strikingly marginalised in Freud's dialogue on war; not only, for instance, sustained engagement with contemporary politics, but also any sense that technology, armaments, the 'war machine', might be a problem *sui generis*. The armaments system is not directly subjected to scrutiny; it provokes no anxiety in itself. It

19 See Chapter 8.

is simply a tool in the hands of its operators (who are the complex forces). Freud identifies the friction between the destructive potential of the machines and the primitive human conduct of war, but poses no 'Frankenstein' scenario here. There is no monster poised between human and machine, pushed into existence and the world only to return upon its maker for some final reckoning. We are offered no sustained consideration of an *inhuman* agency which refuses to submit to human wills. The range of discussion largely concerns human psychology, the control of primitive drives in the sphere of the social, hence the highly charged problem of instinctual 'renunciation' and its discontents. It is perhaps not surprising that Freud's concern and interest should lie in the human rather than the mechanical sphere. My point is not at all to make a prescriptive demand about what the correspondence in 1932 *ought* to have addressed, nor to claim that a concern with the power and dynamics of the war machine would necessarily have been more fruitful or profound; but simply to observe its absence here, whilst elsewhere, as I have shown, the conception of a war machine imbued with a certain 'spirit' or willed energy had become a widespread and powerfully recurrent cultural concern and image.

Freud had begun his reply with an anthropological vignette. Once upon a time in the small human horde, superior muscular strength decided who owned things or whose will should prevail.[20] Muscular strenth and also, more suggestively, an erotic tie, binds the group in its unequal internal relations, Freud had argued in *Group Psychology* (1921).[21] In time, muscular strength was supplanted

20   On the Darwinian context of Freud's theory of the 'primal horde', see Ritvo, *Darwin's Influence on Freud*, p. 100. When Freud presented his primal horde theory for the first time in 1913, he quoted at length from volume II of Darwin's *Descent of Man* (1871). Darwin had deduced from the habits of the higher apes that man too originally lived in comparatively small groups or hordes within which the jealousy of the oldest and strongest male prevented sexual promiscuity. The group was held together by such internal fears and allegiances, and by fear of the enemies that lay beyond the horde.

21   In the chapter 'The group and the primal horde' in *Group Psychology and the Analysis of the Ego*, Freud wrote: 'The uncanny and coercive characteristics of group formation, which are shown in the phenomena of suggestion that accompany them, may therefore with justice be traced back to the fact of their origin from the primal horde. The leader of the group is still the dreaded primal father; the group still wishes to be governed by unrestricted force; it has an extreme passion for authority; in Le Bon's phrase, it has a thirst for obedience. The primal father is the group ideal. Hypnosis has a good claim to being described as a group of two. There remains as a definition for suggestion: a conviction which is not based upon perception and reasoning but upon an erotic tie' (pp. 127–8).

by the use of tools. Victory increasingly belonged to the one with the better weapons or the greater skill in deploying them. This marked the progressive triumph of intellectual superiority over brute muscular strength, although the objective remained the same. As Freud notes in *Why War?*:

> one side or the other was to be compelled to abandon his claim or his objection by the damage inflicted on him and by the crippling of his strength. That purpose was most completely achieved if the victor's violence eliminated his opponent perma- nently — that is to say, killed him. This had two advantages: he could not renew his opposition and his fate deterred others from following his example. (p. 204)

But the killing which marked victory was sometimes suspended. For Freud, the death of the vanquished might be avoided where a slave rather than a corpse was required:

> The intention to kill might be countered by a reflection that the enemy could be employed in performing useful services if he were left alive in an intimidated condition. In that case the victor's violence was content with subjugating him instead of killing him. This was a first beginning of the idea of sparing an enemy's life, but thereafter the victor had to reckon with his defeated opponent's lurking thirst for revenge and sacrificed some of his own security. (*Ibid.*)

The purpose of resisting the desire to kill was apparently utilitarian. In making this point Freud seems partly to echo the Hegelian conflict of the master and the slave; and yet in *The Phenomenology of Spirit* (1807) the function of that non-infliction of death by the victor is recognition, achieved by allowing the vanquished to live on as a slave. The slave acknowledges the master, whose very identity as lord depends upon the acquiescence and recognition of that same vanquished figure. The master's identity is thus precariously dependent on the slave. Coincidentally, the very year that Freud and Einstein's deliberations were published was also to see a startlingly compelling reinterpretation of Hegel in terms of mastery and slavery at an extraordinary Paris seminar.[22]

Freud's speculative history/anthropology moved on to the ques- tion of the union of weaker individuals into a powerful group, and from that process, to the development of laws. The problem of social control and military conflict was located entirely beyond the

---

22   Freud, it would seem, remained oblivious of Kojève's seminar on Hegel. See the discussion in the next section of this chapter, pp. 232 ff.

technology. There was no technological determinism here. Tools were only tools – they had neither minds nor causal force of their own. The same, apparently, could be said of Clausewitz's work, which insists over and over again that war is the tool of the state and therefore always subservient to political will. But then again, had not Clausewitz opened up the possibility of a war machine which disobeys or at least which is held in check only by the most formidable exercise of will? War, like the unconscious, rails against limits, produces friction and disturbs the very possibility of abstract knowledge:

> as soon as difficulties arise – and that must always happen when great results are at stake – then things no longer move on of themselves like a well-oiled machine, the machine itself then begins to offer resistance, and to overcome this the Commander must have a great force of will.[23]

The Clausewitzean discussion focused on what happens within the course of a war, in the friction of its duration. But a further question could also be asked: what happens to history itself when punctuated by war? Is it realised, displaced, shattered, transformed or ultimately little affected by the wars within its passage? An historiographical and moral-philosophical commitment to the view that history was a passage or an unfolding of meaning over time (whether of progress, evolution, degeneration, decadence or punishment) was virtually a *sine qua non* of nineteenth-century historical writing. The question then arose as to what extent war represented a friction within, or a perversion of, the progressive or degenerationist historical trajectory to which commentators almost invariably committed themselves. Much of the subsequent historiography of the First World War has rejoined the question of whether the conflict merely speeded up processes already in motion or whether, on the contrary, it carried world history quite elsewhere; to what extent war determined the future beyond its end according to a logic that preceded its beginning, or alternatively according to a set of determinants that emerged only during the military confrontation itself. Traditionally the First World War has often been seen by social theorists as speeding up processes immanent within pre-war history. As Anthony Giddens writes:

> it still tends to be assumed by sociological authors analysing social development in the current century that, if it had any lasting influence on social organization, the First World War

---

23 Clausewitz, *On War* (Penguin, 1986), p. 145.

merely accelerated trends that were bound to emerge in the long-run in any case. But this view is not at all plausible and could scarcely be countenanced at all if it were not for the powerful grip that endogenous and evolutionary conceptions of change have had in the social sciences.[24]

To describe war as the locomotive of history (to borrow from Trotsky) is ambiguous: it may suggest the dynamic primacy of war, or it may imply that war follows a pre-laid route towards a destination set in advance – the rails are in place, stretching into the distance, before the war train arrives. But as Giddens implies, war both shifts and is shifted by the tracks of history.

If First World War propaganda latched on to the comforting prospect that here was a war to end war, other more troubled texts and voices continued to ask whether a dynamic of perpetual war, or at least an infinite dialectical spiral of war and armistice, had not been set in motion – war as the genie out of the bottle, the riderless horse, the uncontrollable machine, the Frankenstein's monster of the twentieth century.

ANXIETY AND MASTERY

Consider two further texts. The first is from the psychoanalytic work of Melanie Klein and Joan Riviere; the second, notes compiled from a famous lecture programme on Hegel by Alexandre Kojève. Both of these discussions occur in the same decade but they remain quite separate. The theories and contexts are distinct, yet both offer, amongst many other things, powerful and elaborate analyses of the difficult historical process of achieving some security in identity, and could be said to represent oblique but important attempts to address the question, 'why war?'

In 1937 a book entitled *Love, Hate and Reparation* was published. It comprised two lectures written 'in everyday language', by Melanie Klein and Joan Riviere.[25] Since much of my discussion has involved fiction, wartime propaganda and various other representations which explore and emphasise the difference between a national 'us' and 'them', I want first to describe their account of the self as a kind of 'imagined community' (to borrow once again Benedict

24 Giddens, *The Nation State and Violence*, p. 234.
25 These public lectures were delivered under the aegis of the Institute of Psychoanalysis in 1936 at Caxton Hall, Westminster under the general title of 'The Emotional Life of Civilized Men and Women'.

Anderson's term). Riviere usefully sets out a number of the processes at stake in the primary literature that I have already discussed, notably projection and introjection. The articulation of
national difference in propaganda pamphlets and invasion-scare
stories in a period of social, economic and military crisis depended
on the representation of a range of externally threatening forces,
but also upon notions of internal spies, fifth columns, disaffected
agents within the host culture. 'Foreign' forces were found not only
to be startlingly invasive but, still more alarmingly, to meet with a
certain support from an internal world of outcasts, stubborn 'aliens
in our midst'.

Whereas for Freud, in *Beyond the Pleasure Principle*, a fundamental aim of life is death, for Riviere, 'the fundamental aim in life
is to live and to live pleasurably'.[26] This intent is pursued through
an internal security system which attempts to deal with a range of
threats:

> In order to achieve this, each of us tries to deal with and dispose
> of the destructive forces in himself, venting, diverting and fusing
> them in such a way as to obtain the maximum *security* he can in
> life – and pleasures to boot – an aim which we achieve by
> infinitely various, subtle and complicated adaptations. (p. 4)

From birth to death, the particular individual psychic configuration
depends upon the balance between 'love' and 'hate' components,
Riviere contends. The pressure of concern with the international
environment of the late 1930s can be felt both directly and indirectly
in these lectures, with their explorations of security, reparation,
defence and destructive propensities. Riviere occasionally refers
her audience explicitly to current international politics, although
allusions to such contemporary dangers never edge out the insistence
that aggression is fundamental and universal. Riviere argues that an
'instinct of aggression, at any rate for defence is generally recognized as innate in man and most animals'. Present political danger,
the precariousness of the current European situation, draws our
attention to infantile and primitive forces which are not simply
contingent on contemporary politics. Aggression is a radical and
basic element in human psychology; it is quite simply inescapable:
'we have only to look at the international situation, or at the
behaviour in any nursery to see that' (*ibid.*).

The individual and the state strive to obtain 'security' against
'dangerous disintegrating forces of hate and aggression'. Internal

26   Klein and Riviere, *Love, Hate and Reparation*, p. 4.

and external aggression constitute both a precondition for 'survival' and, more disturbingly, a source of pleasure, gratification and guilt. Every ordinary person knows that bad temper, selfishness, meanness, greediness, jealousy and enmity are daily felt and expressed by other people. Moreover we can all sense that aggressive, cruel and selfish impulses in the other are bound up with pleasure, fascination and excitement, whatever capacity one may or may not have to appreciate their existence in oneself. Riviere refers to the savage satisfaction experienced by the speaker of a 'cutting retort', to the thrill and *frisson* of cruel stories, pictures, films, sports, accidents and atrocities. Yet even if we accept or know something about the existence of such factors in ourselves, 'we minimize and underestimate their importance. We do not focus our eyes on them, but keep them in the outer edges of our field of vision.' We blur their experience, comfortingly taking the edge off the alarm we feel in the face of our own half-perceived (or perhaps better our actively unfocused) internal destructive propensities.

Riviere acknowledges the more rational or at least external motives for human hostile emotions such as anger and grievance in the face of deprivation, theft or poverty. But internally too the subject struggles for and against a state of independence whose existence is always partly illusory. Although sharing, waiting and cooperating may increase our 'collective security', they threaten the quest for 'individual security' which, more or less shakily, is founded on unconscious impulses and demands. The re-experiencing across life of the renewed 'anxiety of dependence' re-evokes an infantile world, linked ultimately to the fragmented love and hate components of the infant suckling at the breast.

The infant is made aware of its dependence on the breast precisely because the nipple can be withdrawn against its will, despite its cries and screams:

> He becomes aggressive. He automatically explodes, as it were, with hate and aggressive craving. If he feels emptiness and loneliness, an automatic reaction sets in, which may soon become uncontrollable and overwhelming, an aggressive rage which brings pain and explosive, burning, suffocating, choking bodily sensations; and these in turn cause further feelings of lack, pain and apprehension. (p. 8)

For the newly born, subject and object are not clearly differentiated: 'The baby cannot distinguish between "me" and "not-me"; his own sensations are his world, *the* world to him' (p. 9). Gradually the sense of 'me' as a unity comes into being, existing as such in relation to figures or bits, which the 'I' knows to be 'not I'.

Compared with other animals, the human being endures a long
physical helplessness and dependence, a radical experience of both
love and loss. Tortured with frustrated desire for that which is
sometimes withheld, the infant is brought into contact with something like death, 'a recognition of the *non*-existence of something,
of an overwhelming loss, both in ourselves and in others, as it
seems', whose later psychological consequences are never dissolved.
Early experience:

> brings an *awareness of love* (in the form of desire), and a *recog-
> nition of dependence* (in the form of need), at the same moment
> as, and inextricably bound up with feelings and uncontrollable
> sensations of *pain and threatened destruction* within and without.
> The baby's world is out of control; a strike and an earthquake
> have happened in his world, and this is *because* he loves and
> desires, and such love may bring pain and devastation. Yet he
> cannot control or eradicate his desire or his hate, or his efforts to
> seize and obtain; and the whole crisis destroys his well-being. . . .
> Thus our great need develops for *security* and *safety* against these
> terrible risks and intolerable experiences of privation, insecurity
> and aggressions within and without. (pp. 9–10)

Riviere then goes on to explore the mechanisms of projection and
introjection: processes of pushing outwards as against taking in, or
re-internalising, 'objects'. In the former, 'All painful and unpleasant
sensations or feelings in the mind are by this device automatically
relegated outside oneself; one assumes that they belong elsewhere,
not in oneself' (p. 11). Powerful feelings are channelled outwards,
yet the external menaces return upon the subject in a grave, dirty
and terrifying form precisely because they correspond to such
internally powerful feelings.

> aggression and hate boiling up within are felt in the first
> instance to be uncontrollable; they seem to explode within us,
> and drown and burn and suffocate our bodies in our first experi-
> ence of them. Later in life, too, people can feel like 'bursting'
> with rage, burning to seize what they want, itching to tear out
> someone's eyes (or some other part of them), or choking and
> suffocating with suppressed emotion. (pp. 13–14)

Riviere discusses an intricate balance in normal healthy development which inevitably involves some process of detachment from
the mother:

> Now, some measure of *turning away* from a desired thing in
> order to find it more easily elsewhere is actually another basic

mechanism of our psychological growth. Without some degree of dissatisfaction with our mother's milk and her nipples or with our bottles, we should none of us ever grow up mentally at all. By turning away, and also by subdividing our aims and distributing them elsewhere, the needs both of hunger and of sexual pleasure become detached from the mother. (p. 17)

She also points to more pathological variants, more disturbed turns the subject may take, for instance a violent and sudden wrenching away rather than a gradual process of separation, a pronounced and despairing rejection and withdrawal, a deep and far-reaching deprecation of all much-loved and most-desired things; in short a loss of faith and belief in goodness, or even a miserly and reclusive poisoning of life itself. Conversely: '[The] cardinal phantasy of an ever-bountiful never-failing breast is naturally the defence *par excellence* against the possibility of feelings either of destitution or destructiveness arising in oneself' (p. 24). But such heightened versions are only variants – intensifications, extremist tendencies – within a 'normal' process which can never be simply 'free' of hate, greed and aggression, nor of guilt, anxiety and attempts at reparation.

The psychic division of others into 'good' and 'bad' may serve the function of isolating and localising within an area of relative safety what were originally much less contained and more immediately threatening feelings of hate, revenge and destructiveness. In social discourse too, Riviere adds, a class of persons is produced for whom 'we do not feel any need to love as we do those near us – such as foreigners, or capitalists, or perhaps prostitutes, or a specially hated race – some group whom people feel they may abominate if they like' (p. 14). This constitutes a conduit for aggression and hate, 'just as we provide ourselves with compart-ments and receptacles in our houses which can safely receive the offensive or injurious discharges of our bodies' (p. 15). An imagined community – the subject and its immediate objects – is thus *relatively* safe. Hostility and hate are confined or alienated – located in a variety of forms outside. Cousins, for instance, may be reviled in such a way as to spare siblings (although sometimes vice versa). Or feelings towards people may be shifted and displaced on to things. 'And then we may let loose our hostility and hate on to these plague-spots we have ourselves brought into existence or helped to create' (p. 15).

The account provides a compelling way to begin to engage with some of the processes at work in the constitution of national identities. What it does not attempt, what it cannot resolve perhaps any more than the other models I describe, is when, if ever, in history it may be right not simply to analyse war and its splitting functions, but actually to fight an enemy to the death.

In the material above, it perhaps goes without saying that to acknowledge the propensity for violence and aggression in individuals, the intensity and sometimes the profound hostility of feelings, is not at all to offer blanket condonement of brutal behaviour, nor approbation of war polices for that matter. To speak of individual propensities is a different issue from the ethical question of their enactment. Moreover, the join between a theory of the individual and of society should not be seen as in the least self-evident; indeed, it is fraught with problems, albeit Freud sometimes seems excessively to smooth those out. In my next example, however, the problem of ethics seems more tricky: to what extent are we being offered an account of the unstable violence that produces war, or particular political formations, and to what extent an idealisation of that ferocious and warlike dialectic of slavery and mastery which we have just traced in the fictions of Čapek and Rolland, and in the thought of Freud? Although Alexandre Kojève's account can be read in relation to psychoanalytic ideas, it is decisively different both in tone and content. It uneasily treads the line between the profound analysis and the disturbing reproduction of a certain romanticised drama of conflict and domination, a conception of the relationship between subject and other which leaves scant room for those compensating emphases on 'love', 'gratitude' and 'reparation' which form the other dimension of Klein and Riviere's lectures.

So as a final vignette from the 1930s which so starkly sets out many of the issues of violence and identity present in the selection of stories and theories set out in preceding pages, consider Kojève's remarkable lectures on Hegel in Paris. Whilst in a moment I shall quote directly from Hegel, what I am drawing attention to and reproducing here is a reading of war and conflict made possible by Kojève. This is not intended as an adequate account of Hegel, but as a particularly telling version, produced in and inseparable from the inter-war period and the French 'rediscovery' of Hegelian thought – a philosophy which so often had been anathematised, marginalised or forced into what one commentator has called a 'subterranean "rumination"'' in and beyond the second half of the nineteenth century.[27]

27 Roudinesco, *Jacques Lacan and Co.*, p. 137.

Born a Russian aristocrat, Kojève had become a communist sympathiser and eventually found in Hegel's *Phenomenology* what he took to be the overriding historical importance of the dialectic of the master and the slave. Unexpectedly invited to give lectures from 1933 at the Ecole Pratique des Hautes Etudes, he was to expound his view that Hegel had dazzlingly captured the essential nature of the historical struggle between classes.[28] That process apparently came to an end with Napoleon's conquests and the birth of a modern homogeneous universalist state. Napoleon universalised the idea of the French Revolution, bringing the struggle to a conclusion; Hegel placed this knowledge before human consciousness and thus, putatively, brought philosophy to a close. Subsequent history was a footnote to this climax. 'The Chinese Revolution,' Kojève would eventually declare, 'is nothing but the introduction of the Napoleonic Code into China.'[29] But whatever the supposed achievement of Napoleon, Kojève's reading of the dialectic of the master and the slave emphasises the fundamental and enduring instability of that

28 Kojève began his seminar on Hegel at the Ecole Pratique des Hautes Etudes, Section des Sciences Religieuses in 1933. Over the next few years, Lacan, Bataille, Breton, Queneau and Merleau-Ponty, amongst others, would all attend; see Roth, *Knowing and History*, p. 95 and *passim* for the details of his lectures and for the wider intellectual context. Reading through a Marxist and Heideggerian optic, Kojève would explicitly and forcefully wrest Hegel into a new context of questions about being and knowledge. Kojève's reading can also be situated in relation to a broader cultural fascination with the 'underworld' of bourgeois cultural life and of history itself. His interpretation of Hegel can be related to wider surrealist, psychoanalytic and philosophical projects contemporaneous with it. This was the period of the discovery of Marx's 1844 Manuscripts, the moment when 'alienation' found a central place in French political vocabulary. Roth points out that the French turn to Hegel was also a swerve away from the terms of neo-Kantianism and the totalising ambitions of the natural sciences in relation to philosophy (see Roth, *Knowing and History*). Hegelianism constituted a break with Cartesianism; it provided a philosophy, or better something of a voyage, of the subject in history; *cf.* Poster, *Existential Marxism*, p. 12, which argues that Kojève 'veered' intellectually in the direction of Fascism, by effectively glorifying war and violence as they appear in the *Phenomenology*.
29 Quoted in Lilla, 'The end of philosophy', p. 3. Note that after 1945 Kojève turned to government service in France and away from philosophy teaching. In any event he believed that the major questions of human destiny had been resolved. As Lilla puts it: 'Whether the eventual path of that history was to pass through the Soviet Union or a United States or a European "Latin Empire" was a matter of relative indifference to Kojève, since the final result would be the same. Workers and rulers were to become fat, satisfied cogs in a pacific, classless, post-tyrannical state engineered by sages. All capitalist and communist roads lead to this end. In Kojève's epigram, "Marx est Dieu, Ford est son prophète' (*ibid.*, p. 5).

relationship; the desire of the slave and the dependency of the master preclude petrification.

In Hegel, war is seen to explode the torpor of states. It is indeed Hegel who makes the crucial philosophical statement that the state and war are structurally inseparable. The state is not the alternative to war, but the formation which could only be realised in war. It is in war that a state constitutes itself as subject. Hegel rejects the Kantian endorsement of perpetual peace as in effect a pious abstraction of philosophy from the laws of development immanent within history. For Kant, war has no positive content, but is merely a detour into the irrational with which the community of states would do well to dispense.[30] Such a view, with its echoes of Montesquieu and Rousseau, held out the prospect of a universal pacific Republic. In Kant's view, since not only war but the never-ending and indeed continually increasing preparations for war represented the greatest evil which afflicts civilised nations, it was a duty to strive for 'perpetual peace', for a cosmopolitan society or at least a lasting accommodation, accord and system of arbitration between the nation states. But war, argues Hegel, has philosophical and psychological functions that cannot be wished away any more than can the negativity on which consciousness depends:

> The disparity which exists in consciousness between the 'I' and the substance which is its object is the distinction between them, the *negative* in general. This can be regarded as the *defect* of both, though it is their soul, or that which moves them. That is why some of the ancients conceived the *void* as the principle of motion, for they rightly saw the moving principle as the *negative*, though they did not as yet grasp that the negative is the self. Now, although this negative appears at first as a disparity between the 'I' and its object, it is just as much the disparity of the substance with itself.[31]

It is division which establishes individuality.[32] Indeed the negotiation of war forms part of the necessary process of development for the individual as for the state, the confrontation with its own limit, its own negation. Yet war is also the overcoming of the self.

The willingness of the individual to die in war is taken to disprove not only the case for Kantian cosmopolitanism but also the universalist claims of utilitarianism. 'The ethical moment of war' is the transcendence of selfishness and individualism.[33] War is

30 See Verene, 'Hegel's account of war', p. 176.
31 Hegel, *The Phenomenology of Spirit*, p. 21.
32 See *ibid.*, p. 108.
33 Hegel, *The Philosophy of Right*, p. 209, para. 324.

the crucial founding moment of the state and the invigorating, revitalising force which recurs across its history. In Hegel's arresting and notorious formulation from *The Philosophy of Right* (1821), through war:

> the ethical health of peoples is preserved in their indifference to the stabilisation of finite institutions; just as the blowing of the winds preserves the sea from the foulness which would be the result of a prolonged calm, so also corruption in nations would be the product of a prolonged, let alone 'perpetual' peace.[34]

Human self-consciousness can only be realised through the capture of the desire of the other. '*Self-consciousness achieves its satisfaction only in another self-consciousness.*'[35] It is constituted in a social exchange, or rather a dialectic of appropriation and assimilation.

In war a state confronts itself as well as the enemy; the two confrontations are indeed symbiotic, each party discovering itself via the location of the belligerent opponent. War is in that sense the crucial moment of the coming into being, into identity, of the state as nation. Insight begins in the startling recognition of separation. For Hegel, the clash of nations in war ideally binds each party, turning 'unrest' away from an inner to an outer world: 'Successful wars have checked domestic unrest and consolidated the power of the state at home.'[36] Moreover, war is the vehicle of historical purpose and progress, the means of asserting the mastery of that nation which bears the world's destiny at a given time:

> The nation to which is ascribed a moment of the Idea in the form of a natural principle is entrusted with giving complete effect to it in the advance of the self-developing self-consciousness of the world. This nation is dominant in world history during this one epoch, and it is only once ... that it can make its hour strike. In contrast with its absolute right of being the vehicle of this present stage in the world mind's development, the minds of the other nations are without rights, and they, along with those whose hour has struck already, count no longer in world history. (p. 217, para. 347)

Degeneration, defeat and destruction are inevitable at some point for the individual nation, just as death is written into the individual life. But that defeat does not constitute a general world decline; rather it highlights the displacement of the force of progress:

34 *Ibid.*, p. 210, para. 324.
35 Hegel, *The Phenomenology of Spirit*, p. 110.
36 Hegel, *The Philosophy of Right*, p. 210, para. 324.

A nation makes internal advances; it develops further and is ultimately destroyed. The appropriate categories here are those of cultural development, over-refinement, and degeneration; the latter can be either the product or the cause of the nation's downfall.[37]

When the spirit of the nation has fulfilled its function, its agility and interest flag:

> the nation lives on the borderline between manhood and old age, and enjoys the fruits of its efforts. . . . The natural death of the national spirit may take the form of political stagnation, or of what we call habit. The clock is wound up and runs automatically. . . . Thus both nations and individuals die a natural death. (p. 59)

For Kojève the concept of desire is central to the understanding of both national and individual identity. War raises the stakes very high, thereby dramatically disclosing the nature and power of incompatible desires:

> Desire is what transforms Being, revealed to itself by itself in (true) knowledge, into an 'object' revealed to a 'subject' by a subject different from the object and 'opposed' to it. It is in and by – or better still, as – 'his' Desire that man is formed and is revealed – to himself and to others – as an I, as the I that is essentially different from, and radially opposed to, the non-I. The (human) I is the I of a Desire or of Desire.[38]

This desire which constitutes man as 'human' involves an appropriation and destruction of the object *qua* object. Desire disquiets and moves its bearer to action – an attempt to evacuate desire:

> In contrast to the knowledge that keeps man in a passive quietude, Desire dis-quiets him and moves him to action. Born of Desire, action tends to satisfy it, and can do so only by the 'negation,' the destruction, or at least the transformation, of the desired object: to satisfy hunger, for example, the food must be destroyed or, in any case, transformed. Thus, all action is 'negating.' Far from leaving the given as it is, action destroys it; if not in its being, at least in its given form. And all 'negating-negativity' with respect to the given is necessarily active. (p. 4)

The other is assimilated into a part of the victorious, desiring being. The being that eats, for example, transforms food, an alien

37 Hegel, *Philosophy of World History*, p. 56.
38 Kojève, *Introduction to the Reading of Hegel*, pp. 3–4.

reality, into its own reality. It assimilates, internalises a 'foreign', 'external' reality. And, generally speaking, 'the I of Desire is an emptiness that receives a real positive content only by negating action that satisfies Desire in destroying, transforming, and "assimilating" the desired non-I' (*ibid.*).

This process of appropriating and assimilating the other provides material to fill the initial emptiness of the 'I'. Moreover, the other is forced to recognise the 'I' as desiring, conquering being. The human being is formed in terms of a Desire directed towards another Desire, that is, finally, in terms of a desire for recognition. So the human being can be formed only if at least two of these desires confront one another. Each of the two beings endowed with such a Desire is ready to go to all lengths in pursuit of its satisfaction. It risks its life and the other's in order to impose itself on the other and be recognised. It looks as though their meeting can only be a fight to the death.[39]

But death may, in fact, have to be displaced. In the warring couple, the attainment of the death of the other would preclude the achievement of the identity of the survivor, since there would no longer be an other available to recognise the 'I'. Hence the death struggle is displaced into the dialectic of the master and the slave. Both adversaries must remain alive after the fight. By a supreme, irreducible and unforeseeable act of liberty they must constitute themselves as unequals. The one must be enthralled by the other, fearing, submitting, abandoning the fight for his own recognition. Instead, he must serve and satisfy the desire of the other. He must acknowledge his master and become the master's slave.

Without such a displacement the loser would die and the victor would be robbed of recognition by the vanquished, and hence of achieved identity. Although the master is fixed in his place, the slave is an unstable element, not having desired his state, but having accepted it out of terror. Thus the slave represents change, transcendence, transformation, education. The slave knows what it is to be free and unfree; he wants to return to the freedom he has given up in his acquiescence to the mastery of the victor. Through work the slave achieves a realm of mastery, or potential mastery. The future and hence 'History' belong not to the warlike master, who either dies or preserves himself indefinitely in identity to himself, but to the working slave, who transforms the given world by his work, going beyond himself, and beyond the master. 'If the fear of death, incarnated for the Slave in the person of the warlike

39 See Hegel, *The Phenomenology of Spirit*, p. 113.

Master, is the *sine qua non* of historical progress, it is solely the Slave's work that realizes and perfects it'.[40]

There is, then, a constant possibility that the roles will be de-stabilised. This process characterises states. The early state was constantly challenged with the possibility of assimilation and defeat, since:

> the pagan State is a *human* state only to the extent that it wages perpetual wars for prestige. Now the laws of war, of brute force, are such that the strongest State must little by little swallow up the weaker ones. And the *victorious City* is thus transformed, little by little, into an *Empire* – into the Roman Empire. (p. 62)

To make war on the object in its state of radical difference from the subject is viewed here as the very principle of history and progress. A theory of the passage of civilisations and empires through time is mirrored in a theory of the formation of subjectivity, a dialectic of mastery and slavery which is the precondition for identity.

In this Hegelian scenario of history as recounted by Kojève, there were eventually too few masters to defend the Roman Empire, so the emperor resorted to mercenaries. As a result the citizens of the city were no longer obliged to make war themselves and eventually became incapable of the virtues of the warrior. Consequently, the emperor was able to subjugate them as his patrimony. The former citizens became slaves of the sovereign because, psychologically, they had in fact already become slaves. For in effect, to be a master is to fight, to risk one's life. Hence the citizens who no longer wage war cease to be masters, and that is why they become slaves of the Roman emperor. And that is also why they accept the *ideology* of their slaves: first Stoicism, then Scepticism, and – finally – Christianity (see *ibid.*, p. 63). It is in this condition of pseudo-slave that, for Hegel, the possibility of a Bourgeois is born.

The behaviour of the state in some respects appears to be like a kind of vast but, by definition, unsatisfiable digestive system. The state feeds and digests but its desire is irreducible to a need or a want for available nourishment. It is this irreducibility of human and state desire to animal need for survival which defines man as human:

> Man 'feeds' on Desires as an animal feeds on real things. But in the nature of desire and human reason, man is fundamentally separated from the world of the animals, not simply an animal with those features added. The human I, realized by the active

40   Kojève, *Introduction to the Reading of Hegel*, p. 23.

satisfaction of its Desires, is as much a function of its 'food' as the body of an animal is of its food. (p. 6)

For there to be Self-Consciousness, Desire must therefore be directed toward a non-natural object, toward something that goes beyond the given reality. Now, the only thing that goes beyond the given reality is Desire itself. (p. 5)

But there is still a further sense in which war defines man's desire as 'human'. War is the moment *par excellence* where the subject risks death, thereby transcending the supreme animal concern for self-preservation. The fundamental difference between the human and the animal is the discrepancy between the nature of their desires. The supreme value for an animal is its animal existence, its desire to preserve its life. Human desire, therefore, must overcome this will for preservation. In other words, man's humanity fully emerges only if he risks his (animal) life for the sake of his human desire. Through this risk, human reality is created, realised, verified. 'And that is why to speak of the "origin" of Self-Consciousness is necessarily to speak of the risk of life (for an essentially nonvital end)' (p. 7).

'ALL MY LIBIDO...'

I have sought to map out a history of representations of war and national identity which owes as much to cultural fantasy and desire, to anxiety and the quest for its mastery, as to material military threat. These readings from the 1930s both rejoin many of the earlier debates and provide ways of thinking about them. I have tried to suggest the complexity of the domestic political anxiety at stake in, say, late nineteenth-century war alarmism, and the uses of war talk in producing, consolidating or reassuring a putatively unified community of readers and political subjects of their difference from an actual or prospective enemy. Psychoanalysis clearly offers a language with which to think about such texts.

But what I have been touching on simultaneously is the question of how such texts could also be related to the work of Freud and other analysts. How might we begin to produce an historical account of psychoanalytic writing on war? What other projects, attacks, speculations, sciences, fictions would we link with it? What wider context do we need to provide? Is psychoanalytic interest in war primarily to be understood in relation to other twentieth-century experiences and models, a world of meaning and disturbance perhaps only really born between 1914 and 1918? Or should we link Freud on war with older philosophical continuities and theoretical affiliations?

Psychoanalytic thought on war involves an exploration of the interaction of a social phenomenon and the individual psyche. Although the very viability of these two terms placed in opposition to one another, the very idea that these categories pre-exist their interaction, is itself radically called into question.[41] In many psychoanalytic accounts, war is understood as projection or mobilisation of the inner world. But there has also been some consideration of the ways in which wars and the threat of wars affect the mind through their traumatic invasion of conventional internal psychic defence systems. Whilst common human features of such disturbance could perhaps be adduced and explored, analysis would always inevitably also show the infinitely different ways in which individual patients in the long run experience external war shock: its injuries, disturbances, deaths, catastrophes. The multi-faceted experience of war is seen in psychoanalysis to interact with some-

41 In other words the contrast made here between the individual and the organised community, which is certainly derived from many of Freud's own pronouncements, is at the same time inadequate to the full complexity of the interaction between these terms which emerges in Freud's account. In an extremely intricate study of the relation between concepts of the individual and the group, Mikkel Borch-Jacobsen shows that Freud undercuts the traditional 'social psychology' notion of a relationship between a (pre-given) individual and a crowd. He makes the important point that the opposition between 'individual psychology' and 'social psychology' is itself found in Freud to be secondary to the more fundamental opposition between 'narcissistic psychic acts' and 'social psychic acts', although again each of those terms could be further questioned. The category of the 'social bond' is transformed into the primary issue of the subject's relation to others, its formation through others. The point is that 'my' relation to others, to objects, is itself basic to the very constitution of 'me' as subject, not simply some external relationship an already formed 'individual' has with figures or crowds in the world. Borch-Jacobsen's study presents particular difficulties and problems which I cannot claim to have grasped or worked through in the present study. But his formulation makes clear the inadequacy of any simple dichotomy between the concept of an individual and an external group in the Freudian account: 'Thus Freud is not emulating the sociologists by describing definite, established social relations; rather, he is seeking to work back towards the very possibility of social relations; a possibility that such relations necessarily presuppose. No sociality can be instituted before the opening towards others comes about, and whoever holds the key to that "before" also holds the key to everything that comes "after"' (*The Freudian Subject*, p. 131). The point is not that psychoanalysis should inform a pre-existing sociology of the crowd or group (Le Bon, McDougall), but that it reconceives 'the social' at a more profound, internal level; it describes what Borch-Jacobsen calls an 'archi-sociology'. This no doubt has implications for an understanding of war which would need to be further clarified.

thing the subject has known, wished and dreaded unconsciously. For Freud and his followers, this cannot be explained away as 'the mass psychology of Fascism' or any other delimited historical phenomenon. Aggression is fundamental; it may be modified but cannot be abolished by political decree or historical transformation.

To consider Freud on war is to focus not only on explicit discussion of national conflict, nor simply to cross-reference war to related papers on religion, civilisation, death, crowds and so forth, but to explore a range of other implications. The pressure of war in Freud's work can be felt at many levels. We have Freud's declarations, pronouncements, insights, but also, inextricably, his metaphors, the war imagery that appears across his writing – from talk of city citadels and the 'battlefield of transference' to his affectionate reference to his *History of the Psychoanalytic Movement* as the 'bomb'.[42] Freud drew attention to this domain of war and words, the slips of the tongue which he recorded again and again in the First World War in which two of his sons were fighting – that vast conflict about whose début he confided: 'All my libido is given to Austro-Hungary.'[43] Close to death in exile in London in the late 1930s, Freud would be asked whether he thought the war against the Nazis would be the final war, the war to end war (as had forlornly been argued one world war earlier) to which he wryly replied: 'Anyhow it is my last war.'[44]

Freud endured interminable intellectual, physical and institutional conflicts. Opponents have always tended to bristle with weapon metaphors against him, sometimes even in the very process of diagnosing the belligerence of the psychoanalytic movement. The tendency continues. If we turn to the highly reductivist critique of psychoanalysis offered by Eysenck we find the language heavy with military terms, references to the arms which will be wielded against Freud. Eysenck's *The Decline and Fall of the Freudian Empire* views psychoanalysis as maimed by the distorted histories written by its own 'camp-followers',[45] vitiated by ill-informed appraisals

---

42 See Gay, *Freud*, pp. 301, 241. Jacquline Rose has recently pointed to the extraordinary imagery of military conflict used (and commented upon) in the 'Controversial Discussions' during the Second World War. See Rose, *Why War? Psychoanalysis, Politics and the Return to Melanie Klein*.

43 Jones, *Sigmund Freud: Life and Work*, vol. II, p. 192. Cf. Freud's acknowledgement to Abraham: 'for the first time for thirty years I feel myself to be an Austrian and feel like giving this not very hopeful Empire another chance' (Letter of 26 July 1914, Freud and Abraham, *A Psychoanalytic Dialogue*, p. 186).

44 Quoted in Jones, *Sigmund Freud*, vol. III, p. 262.

45 Eysenck, *Decline and Fall*, p. 7.

'written more as weapons in a war of propaganda than objective assessments' and as currently characterised by 'internecine war' amongst its various groups, espeically in New York (p. 18). Freud was supposedly a self-deceiving general who also successfully duped the world: 'Supported by his followers he was quite successful in impressing the world with this completely untruthful picture of himself and his battles' (p. 25). Eysenck is apparently convinced of his own attempt at scientific neutrality, at non-interference in 'facts': 'I have tried to deal with ascertained facts, and to add as little comment and interpretation as possible' (p. 7). Yet his own highly contentious 'campaign' proceeds on the basis of various unremarked military expressions: 'The second great weapon in the scientist's *armamentarium*,' he observes, 'is the putting forward of alternative hypotheses' (p. 13).

Above all in the early days of psychoanalysis, the beleaguered position of Freud's work was powerfully real in European medicine, psychiatry and society even if on occasion this was also powerfully amplified by Freudian historiography.[46] This adversarial position was not without importance to the very project of psychoanalysis, to the culture of its intellectual radicalism.[47] On the other hand, Freud took pride in the fact that despite the difficulties of communication, the international bonds of psychoanalysis had not collapsed in 1914, like some other 'Internationals'.[48] Nevertheless, the psychoanalytic movement had suffered its own internal battles and breaks – Adler, Steckel, Jung. As Roazen comments: 'Freud's system of thought reflected his fighting stance; he used military language and the image of warfare throughout – attack, defence, struggle, enemy, resistance, supplies, triumph, conquest, fight.'[49]

46 See Sulloway, *Freud, Biologist of the Mind.*
47 Elizabeth Roudinesco's extraordinary account of the history of psycho-analysis in France is entitled, *La Bataille de cent ans.* Sometimes poignant, but often horrific and grotesque, this meticulously told story draws through-out on the metaphors of violence: 'hot wars', 'cold wars', 'battle', an 'army of barons'.
48 As Freud wrote in 1925: 'The World War which broke up so many other organisations, could do nothing against our "International". The first meeting after the war took place in 1920, at The Hague, on neutral ground. . . . I believe this was the first occasion in a ruined world on which Englishmen and Germans sat at the same table for the friendly discussion of scientific interests. Both in Germany and in the countries of Western Europe the war had actually stimulated interest in psychoanalysis' ('An autobiographical study', p. 54). But *cf.* his remark in 1915: 'of all my colleagues I see only Ferenczi standing out against the military influence and sticking to the group' (Freud and Andreas-Salomé, *Letters*, p. 32: 30 July 1915).
49 Roazen, *Freud and his Followers*, p. 194.

War provided decisive metaphors for the intellectually and professionally besieged world that Freud experienced and portrayed, as here, for instance:

> I am living, as my brother says, in my primitive trench; I speculate and write, and after hard battles have got through the first line of riddles and difficulties. Anxiety, hysteria, and paranoia have capitulated. We shall now see how far the successes can be carried forward.[50]

In fact, Freud was to prove extremely productive during the war, 'as often happened,' remarks Jones, 'when he felt in poor health or low spirits'.[51] Psychoanalysis itself both suffered and flourished in the real war, as it had in the atmosphere of conflict and incomprehension that had surrounded its own inception. Various journals of sociology and psychiatry began to take an interest in 'the talking cure' in the face of a crisis whose barely nameable name was male hysteria.[52] From a military point of view here was an unprecedented catastrophe in management afflicting both officers and troops, indeed officers more than troops, and which defied all conventional nineteenth-century models of madness, challenging the image of strength, nobility and resolve widely attached to the volunteers of the First World War. The sense of crisis aroused by shellshock owed a great deal to the expectations of manliness fostered by the war propagandists and to the wider presumption that mental illness was an index to the moral state of the nation – as though the sum of individual physical and psychological conditions formed a collective mind and body.

THE PSYCHOPATHOLOGY OF EVERYDAY DEATH

Freud initially reacted to war in 1914 with a certain enthusiasm combined with relief that his own sons apparently would not be called up.[53] Ernest Jones insists on the prosaic nature of many of

---

50 Quoted in the introduction to Freud, *Phylogenetic Fantasy*, p. 77.
51 Jones, *Sigmund Freud*, vol. II, p. 199.
52 *Cf.* Eder, *War Shock*, p. 144. In this account, 'war shock' was deemed to be hysteria occurring in a person free of hereditary or personal psycho-neurotic antecedents, a function of 'the peculiar and terrible mental strain of modern war-conditions' acting upon the sensitive mind.
53 'We are overjoyed that none of our sons or sons-in-law is personally affected, and yet really ashamed of this in view of the multitude of victims all round us' (Letter of 29 July 1914, Freud and Abraham, *A Psycho-Analytic Dialogue*, p. 187). The Austrian authorities had rejected two of them and exempted

Freud's political responses, even to the point of declaring that 'In his judgement of political events, Freud was neither more nor less perspicacious than another man. He followed them, but had no special interest in them unless they impinged on the progress of his own work. 1914 was the first time they did so.' Whilst Freud's political attitudes had hitherto struck his biographer as somewhat conventional and limited, Jones admits that he *is* surprised by the professor's attitude to the war:

> Freud's immediate response to the declaration of war was an unexpected one. One would have supposed that a pacific *savant* of fifty-eight would have greeted it with simple horror, as so many did. On the contrary, his first response was rather one of youthful enthusiasm, apparently a reawakening of the military ardours of his boyhood. . . . . He was quite carried away, could not think of any work, and spent his time discussing the events of the day with his brother Alexander. . . . He was excitable, irritable, and made slips of the tongue all day long.[54]

Freud's excitement about war may be suprising to Jones, but it was hardly exceptional within the European intelligentsia at large. That involvement can no doubt be interpreted in different ways, but surely what is important here is not simply to remark on how Freud was caught up like so many of the rest in war enthusiasm, but to see the implications of Freud's (fluctuating) acknowledgement of this process of excitement and disturbance. Psychoanalysis after all was to come increasingly to recognise the implication of the observer in the observed, as it was to confront rather than efface the issue of the analyst's counter-transference.[55] Indeed an exploration of the

---

the third. But Freud's eldest son Martin volunteered in August and wrote enthusiastically to his father in anticipation of military action. Oliver, Freud's second son, became an engineer in 1915, whilst Ernst had volunteered by October 1914. None was killed in the war. See Gay, *Freud*, p. 352. Freud's concern for his sons' fate in the war runs through his letters to Lou Andreas-Salomé; see Freud and Andreas-Salomé, *Letters*.

54 Jones, *Sigmund Freud*, vol. II, pp. 189–92.
55 Note that in Freud's early discussion of the term the aim seems to be to surpass rather than (as in later developments) to recognise and use the counter-transference. Nevertheless the concept of the analyst's own unconscious involvement in the encounter is clearly there in formulations such as the following: 'We have become aware of the "counter-transference", which arises in him [the physician] as a result of the patient's influence on his unconscious feelings, and we are almost inclined to insist that he shall recognize this counter-transference in himself and overcome it. Now that a considerable number of people are practising psycho-analysis and exchanging their observations with one another, we have noticed that no psycho-analyst

question 'why war?', whether avowedly scientific or not, cannot simply disavow identification and implication, cannot take up some neutral stance without manifesting a symptomatic denial, the denial to which, it could even be said, Freud returned when he wrote to Einstein of the pacifist's constitutional intolerance of war. Earlier, Freud had acknowledged the confusion the war had provoked in him – even an unconscious hostility to his beloved Italy:

When war broke out with Italy in 1915 I was able to make the observation upon myself that a whole quantity of Italian place names which at ordinary times were readily available to me had suddenly been withdrawn from my memory. Like so many Germans I had made it my habit to spend a part of my holidays on Italian soil, and I could not doubt that this large-scale forgetting of names was the expression of an understandable hostility to Italy which had now replaced my former partiality.[56]

But beyond his initial identification with Austro-Hungary in 1914, sadness and disenchantment characterise many of Freud's utterances on the First World War. 'Now I am more isolated from the world than ever, and expect to be so later too as the result of the war,' he wrote to Sandor Ferenczi, Hungarian analyst and close collaborator, on 15 December 1914. 'Germany has not earned my sympathy as an analyst and as for our common Fatherland the less said the better'.[57] Or as he wrote to Lou Andreas-Salomé:

I do not doubt that mankind will survive even this war, but I know for certain that for me and my contemporaries the world

goes further than his own complexes and internal resistances permit; and we consequently require that he shall begin his activity with a self-analysis and continually carry it deeper while he is making his observations on his patients. Anyone who fails to produce results in a self-analysis of this kind may at once give up any idea of being able to treat patients by analysis' ('The future prospects of psycho-analytic therapy', pp. 144–5).

56 Freud, *Psychopathology of Everyday Life*, pp. 33–4. This paragraph was added to the text in 1917. Many others acknowledged and reflected on this too, of course. Even the pacifist Bertrand Russell could not conscientiously object to national identifications altogether, but admitted: 'I was myself tortured by patriotism. The successes of the Germans before the battle of the Marne were horrible to me. I desired the defeat of Germany as ardently as any retired colonel. Love of England is very nearly the strongest emotion I possess, and in appearing to set it aside at such a moment I was making a very difficult renunciation... I hardly supposed that much good would come of opposing the war, but I felt that for the honour of human nature those who were not swept off their feet should show that they stood firm' (Russell, *Autobiography*, vol. II, pp. 17–18).

57 Quoted in Jones, *Sigmund Freud*, vol. II, p. 199.

will never again be a happy place. It is too hideous. And the saddest thing about it is that it is exactly the way we should have expected people to behave from our knowledge of psychoanalysis. . . . My secret conclusion has always been: since we can only regard the highest present civilization as burdened with an enormous hypocrisy, it follows that we are organically unfitted for it. We have to abdicate, and the Great Unknown, He or It, lurking behind Fate, will sometimes repeat this experiment with another race.[58]

Freud remarked caustically on the temptation to blame the war on a few 'evil' individuals, when in fact it bore witness to a dynamic of regression in which perhaps *everyone* was caught up, disclosing something of the mental constitution and not simply the aberration of a minority:

And now turn your eyes away from individuals and consider the Great War which is still laying Europe waste. Think of the vast amount of brutality, cruelty and lies which are able to spread over the civilized world. Do you really believe that a handful of ambitious and deluding men without conscience could have succeeded in unleashing all these evil spirits if their millions of followers did not share their guilt? Do you venture, in such circumstances, to break a lance on behalf of the exclusion of evil from the mental constitution of mankind?[59]

The war, as he put it, 'brought us all such constant and protracted preoccupations'; confusions and parapraxes were to be expected. The later editions of *The Psychopathology of Everyday Life* record not only patients' and acquaintances' slips of the tongue, but also his own.[60] We cannot simply dissociate ourselves, Freud suggests, from the thrill of war; it infiltrates and impinges upon the mind, but also evokes something already present. It is a crossroads of the social, the political and the unconscious. War is a particular kind of organisation, but it affects *all* institutions. In his account of psycho-

58 Freud, letter to Lou Andreas-Salomé in Freud and Andreas-Salomé, *Letters*, p. 21: 25 Nov. 1914.
59 *Introductory Lectures*, pt II, 'Dreams', p. 146.
60 For instance: 'One day I picked up a mid-day or evening paper and saw in large print: "*Der Friede von Görz* [The Peace of Gorizia]." But no, all it said was "*Die Feinde vor Görz* [The enemy before Gorizia]." It is easy for someone who has two sons fighting at the very time in that theatre of operations to make such a mistake in reading' (*Psychopathology of Everyday Life*, p. 113). Cf. Freud's reference to the date of the Budapest Congress as 1818 rather than 1918 in 'Memorandum on the electrical treatment of war neurotics', p. 215.

analytic history he discloses the warlike manoeuvres that have taken place around his own movement, although on this occasion he distances himself from such motives:

> But I must add at once that it has never occurred to me to pour contempt upon the opponents of psycho-analysis merely because they were opponents – apart from the few unworthy individuals, the adventurers and profiteers who are always to be found on both sides in time of war.[61]

He insists that psychoanalysis has long known the state of war, its battlefield picked over by adventurers and profiteers. Yet was he not (in his own self-perception) a *conquistador*, his heroes, Hannibal, Napoleon, Alexander the Great, Oliver Cromwell? Freud fluctuated between the desire to protect and lead a movement, and the stoical recognition, in line with his own theories, that the author of a body of thought and an institution could never be in full mastery of its developments.

The potential psychoanalyst today, warns Freud in the *Introductory Lectures* (1916–17):

> would find himself in a society which did not understand his efforts, which regarded him with distrust and hostility, and unleashed upon him all the evil spirits lurking within it. And the phenomena accompanying the war that is now raging in Europe will perhaps give you some notion of what regions of these evil spirits there may be.[62]

On various occasions in his work psychoanalysis is viewed as victim in a war of several fronts, caught between different enemies, repulsing the assault on one flank only to be attacked from another, too often caught in a shadowy no man's land or a dangerous borderland between, say, medico-psychiatry and the occult, or between the alliance and the treachery of various hitherto staunch lieutenants, like Jung and Adler:

> Scarcely have we triumphantly repulsed two attacks – one of which sought to deny once more what we had brought to light and only offered us in exchange the theme of disavowal, while the other tried to persuade us that we had mistaken the nature of what we had found and might with advantage take something else in its place – scarcely, then, do we feel ourselves safe from these enemies, when another peril has arisen.[63]

61 'On the history of the psychoanalytic movement', pp. 38–9.
62 *Introductory Lectures on Psycho-analysis*, p. 16.
63 'Psycho-analysis and telepathy', p. 177. The new adversary was the 'occult'.

Or again:

> We have heard during the war of people who stood half-way
> between two hostile nations, belonging to one by birth and the
> other by choice and domicile; it was their fate to be treated as
> enemies first by one side and then, if they were lucky enough to
> escape, by the other. Such might equally be the fate of psycho-
> analysis. (*Ibid.*, p. 180)

Psychoanalytic theory and influence were affected by the war
although we should not take 1914–18 as in any simple sense the
origin of the work on death forces or aggression. Nor should we
accept Freud's insistence that war had revealed nothing new – 'I
cannot be an optimist,' he wrote to Lou Andreas-Salomé in 1915,
'and I believe I differ from the pessimists only in that wicked,
stupid, senseless things don't upset me, because I have accepted
them from the beginning as part of what the world is made of.'[64]
The war, as Freud noted in the introduction to a psychoanalytic
study of shellshock, 'was not without an important influence on the
spread of psychoanalysis', because medical men 'who had hitherto
held back from any approach to psychoanalytic theory were
brought into close contact with them when in the course of
their duty as army doctors they were obliged to deal with war
neuroses'.[65] The book had arisen from contributions to the fifth
International Psychoanalytical Congress held in Budapest in late
September 1918. A symposium had been held on 'The Psychoanalysis of War Neuroses'. Papers were presented by Ferenczi,
Abraham and Simmel. These three contributions, together with
another by Jones on the same topic (which had been read in
London before the Royal Society of Medicine on 9 April 1918),
were published a year later in a small volume, the first to be issued
by the newly founded International Psychoanalytical Publishing
House (Internationaler Psychoanalytischer Verlag). Freud provided
a short introduction, noting how official observers from the highest
quarters of the Central European Powers were present as observers
at the Budapest Congress.[66] In Freud's words,

> The hopeful result of this first contact was that the establishment
> of psychoanalytic Centres was promised, at which analytically

64 Freud, letter to Lou Andreas-Salomé, in Freud and Andreas-Salomé, *Letters*,
    p. 33: 30 July 1915.
65 'Introduction to *Psycho-Analysis and the War Neuroses*', p. 207.
66 See Gay, *Freud*, p. 376. For a comment on the German War Ministry's
    interest in the Budapest Conference, see Freud and Andreas-Salomé, *Letters*,
    p. 83: 4 Oct. 1918.

trained physicians would have leisure and opportunity for studying the nature of these puzzling disorders [the war neuroses] and the therapeutic effect exercised on them by psychoanalysis.[67]

Before these proposals could be put into effect, however, 'the war came to an end, the state organisations collapsed and interest in the war neuroses gave place to other concerns' (p. 207). Moreover 'the greater number of the neurotic disturbances brought about by the war simultaneously vanished. The opportunity for a thorough investigation of these affections was thus unluckily lost – though, we must add, the early recurrence of such an opportunity is not a thing to be desired' (*ibid.*). Freud had even written gloomily to Ferenczi: 'Our psychoanalysis has had bad luck. No sooner had it begun to interest the world because of the war neuroses than the war comes to an end.'[68] On the other hand Karl Abraham wrote to Freud of his scepticism about the value of any *rapprochement* between medico-psychiatry and psychoanalysis occasioned by shell-shock: 'I did not like the idea that psychoanalysis should suddenly become fashionable because of purely practical considerations.'[69]

Shellshock *was* a practical consideration, but also a tragedy of human suffering and a profound challenge to orthodox medical approaches.[70] For conventional medico-psychiatry, the First World War disturbances presented real diagnostic difficulties: how to make sense of this 'no man's land' of illness, which seemed to negate commonly held beliefs about valour and masculinity, and to defy the prevailing organic models of insanity and its aetiology? The idea that the shellshocked were all hereditary degenerates or that their condition could be put down to the commotional effects of exploding shells on the central nervous system proved increasingly unsustainable. Yet shellshock could not be explained away as malingering. It blurred the distinctions between neurosis and insanity – and it was a crisis on a massive scale. According to one account in 1916, shellshock cases constituted up to 40 per cent of the casualties from heavy fighting zones; more alarmingly still, officers seemed especially prone to it. Army statistics revealed that

---

67 See 'Introduction to *Psycho-Analysis and the War Neuroses*', p. 207.
68 Quoted in Hoffman 'War, revolution and psychoanalysis', p. 251. Cf. Freud's remark *à propos* of Simmel's work on shellshock: 'the German medical authorities in the war zones are at last beginning to make use of analysis and are able to report well of it' (Freud and Andreas-Salomé, *Letters*, p. 67: 22 Nov. 1917).
69 Letter of 2 December 1918, Freud and Abraham, *A Psycho-Analytic Dialogue*, pp. 279–80.
70 See Stone, 'Shellshock and the psychologists'.

officers were more than twice as likely to suffer from mental break-down on the battlefield as men of the ranks. Here, then, was 'a mass epidemic of mental disorders of an extent hitherto unknown in war', and this related, as Martin Stone puts it, to the fact of 'a modern industrial mode of warfare that entailed on the one hand mass production in munitions factories and on the other, the deployment of barely trained civilian armies on the battlefield' (p. 248).

In the later nineteenth century, increasing casualty rates from industrial and train accidents had brought the problem of delayed injury, industrial malingering and financial compensation into focus. The condition of 'railway spine' for instance might not be immediately manifest, but was primarily viewed as an organic disturbance precipitated by the shock of an accident. But as with the later discussion of shellshock, physicalist explanations were gradually abandoned in favour of psychological models. New dis-cussions of risk, compensation and insurance clustered around the problem of industrial injury, resulting in government legislation in 1897. A whole array of industrial accidents and traumatically induced conditions – defects of speech, tics, paralysis and so forth – preoccupied specialist commentators before the war.[71] The terms of the shellshock debate and the aetiological dilemma were in many respects in place before 1914.

Whilst it must also be emphasised that the vast majority of shell-shock patients continued to be treated by conventional means,[72] what opened up in the fog of war was an increasingly misty space for diagnosis. In 1916, the *Lancet*, for instance, published a piece entitled 'Neurasthenia and shellshock' which rejected popular mis-conceptions about the polarisation of madness and health; it focused instead on 'the borderline cases'.[73] Frontiers in medicine, as in political geography, had to be viewed as arbitrary: 'In medicine there is a neutral zone, a no-man's land . . . which really defies

71 I am indebted here to Dr Michael Clarke for an illuminating talk on shell-shock and its pre-war industrial context which accompanied the presentation of previously unseen film archives on the treatment of First World War patients, given under the auspices of the Wellcome Institute at the Institute of Historical Research, London in December 1991.

72 Even figures like W.H.R. Rivers restricted his advocacy of Freud to the claim that a minority of cases could benefit from psychoanalysis; the crisis of shellshock enhanced psychoanalytic claims as a cure rather than as primarily a psychological model or method of scientific inquiry; nevertheless it contributed to the intensifying public interest in Freud during the 1920s; see Rapp, 'The early discovery of Freud by the British general educated public'.

73 *The Lancet*, 18 March 1916, p. 627.

definition' (p. 627). Many shellshock cases occupied this 'nebulous zone'. Indeed with the war, the *Lancet* claimed, we want 'a brand new dictionary' of terms (pp. 627–8). In an article entitled 'Shell blindness' (subtitled 'The problem of wounds to consciousness'), *The Times*, a year earlier, had also called for sympathy and open minds. Shell blindness was not necessarily the result of direct injury, but the consequence of some more obscure concussion in which the man became automaton-like: 'Consciousness is lost for a variable time but often not so far as to prevent automatic movements, so that the man may walk in a dazed condition to the dressing station.'[74] The blindness was usually found to be temporary, and a gradual recovery occurred. The eye to recover last, however, was the shooting eye. Some of the patients admitted candidly to being 'in a blue funk', but shamming was apparently not an adequate answer:

> The problem is psychological and clearly demands a most careful and minute study of the evolution of character. It would seem that as a result of severe and sudden shock the conscious mind, with its high attributes of control and determination, is thrown out of action; the 'subconscious mind' supervenes. This subconscious mind, about which so much has been written of late, is a kind of storehouse of forgotten and unremembered things – events and ideas which belong to the past of the race as well as to that of the individual, but of the presence of which within him the individual is scarcely conscious, or not at all conscious, as the case may be. The highest centres in his brain cease to function, powerful primeval instincts resume sway and a 'block' occurs between the mechanism and the perception of sight. (p. 6)[75]

Psychoanalytic explanations of shellshock stressed the primary gain from being ill, the importance of unconscious instinctual impulses, the psychogenetic origins of symptoms, and (most controversially) the link between war neuroses and sexuality:

> In traumatic and war neuroses the human ego is defending itself from a danger which threatens it from without or which is

74 *The Times*, 8 April 1915, p. 6.
75 Note, however, that the War Office Committee of Inquiry into Shellshock under the chairmanship of Lord Southborough in 1922 entertained but then rejected Freud's therapy, or at least the 'sanitised' versions they had been offered by Head and Rivers. The committee declared that Jews, the Irish and the working classes were more likely to break down, as were 'artistic types' and 'imaginative city-dwellers' and other such 'highly strung' people; see Bogacz, 'War neurosis and cultural change in England'.

embodied in a shape assumed by the ego itself. In the transference neuroses of peace the enemy from which the ego is defending itself is actually the libido, whose demands seem to it to be menacing. In both cases the ego is afraid of being damaged – in the latter case by the libido and in the former by external violence. It might, indeed, be said that in the case of the war neuroses, in contrast to the pure traumatic neuroses and in approximation to the transference neuroses, what is feared is nevertheless an internal enemy.[76]

Many commentators, of course, continued to berate psycho-analysis as unscientific, vague or therapeutically useless. To some army officers it seemed, not without some reason, that psycho-analysis was potentially subversive, damaging to the uncritical allegiance required in war. Freud described a disciple of psycho-analysis, a German medical officer who managed to exercise a powerful influence on one of his patients. Questioned by other officers, he explained something of psychoanalytic theory to his colleagues:

> All went well for a while: but when he spoke to his audience about the Oedipus complex, one of his superiors rose, declared he did not believe it, that it was a vile act on the part of the lecturer to speak of such things to them, honest men who were fighting for their country and fathers of a family, and that he forbade the continuance of the lectures.... The analyst got himself trans-ferred to another part of the front. It seems to me a bad thing, however, if a German victory requires that science shall be 'organised' in this way.[77]

In England too, suspicion did not evaporate in the face of the new interest. When Arthur Brock for instance cited various methods of dealing psychologically with neurasthenia in an article which appeared in the *Sociological Review* in 1918 he referred dis-paragingly to psychoanalysis as a flimsy method of 'worming out of the patient's subconsciousness ... suppressed wishes'. It was, he argued, reminiscent of the 'obstetric' procedure of Socrates, who taught men by 'bringing their thoughts to birth'.[78] Brock seemed unsure whether the analyst might be delivering or actually impregnating the thoughts which emerged, but either way he was doubtful of the benefit. He spoke of 'the intensified psychology of war' and the war-neurasthenic as 'an extreme instance of the

76 Freud, 'Introduction to *Psycho-Analysis and the War Neuroses*', p. 210.
77 *Introductory Lectures on Psycho-Analysis*, vol. XVI, p. 330.
78 Brock, 'The re-education of the adult', p. 30.

chronically fatigued person so common in our modern world – the *vaincu de la vie*' (p. 26). Like many of his colleagues, Brock was reluctant to abandon an organic and evolutionist model, but unclear how to grasp the psychic processes at stake in shellshock. He rejected psychoanalysis but was able only to gesture vaguely at the idea that the neurasthenic manifested a reversal of evolution (p. 39).

Other psychiatrists, however, overcame their initial suspicion of psychoanalysis, and now rejected the 'knee-jerk' disparaging view. In a famous article in the *Lancet* in 1917, W.H.R. Rivers, director of the Craiglockhart Hospital for shellshock victims, declared that:

> It is a wonderful turn of fate that just as Freud's theory of the unconscious and the method of psycho-analysis founded upon it should be so hotly discussed, there should have occurred events which have produced on an enormous scale just those conditions of paralysis and contracture, phobia and obsession, which the theory was especially designed to explain.[79]

Rivers noted the 'hotbed of prejudice and misunderstanding' stirred by Freud's work. This was, he admitted, a conflict produced by the 'extravagance of Freud's adherents' and the 'rancour of their opponents' (p. 912). He declared himself unable to avoid this 'maelstrom of medical controversy' in the context of the war, since Freud's theory was so widely applicable and so relevant to the physician, the teacher, the politician, the moralist and the sociologist. The Viennese provenance of the theory of psycho-analysis should not be allowed to short-circuit the debate. There could be no simple enemy territory in the sphere of ideas. Or as Rivers inquired, 'Are we to reject a helping hand with contumely because it sometimes leads us to discover unpleasant aspects of human nature and because it comes from Vienna?' (p. 914). It would be useful, he suggested, if the reader first of all admitted that unconscious experience ('a vast body of experience of which they have no manifest memory') had at least some effects. Psycho-analysis, he argued, was building on a fund of common sense. Was it not a commonplace today to admit the importance of heredity? Had not such writers as Samuel Butler earlier described heredity as a species of memory? Rivers's strategy was to 'domesticate' Freud – to show that his views were not totally outlandish. Secondly, he stressed the utility of Freudian conceptions of mental conflict, dissociation and above all of forgetting as an active process.

79   Rivers, 'Freud's psychology of the unconscious', p. 913.

But Rivers offered a very partial endorsement. He suggested that shellshock disproved the importance of sexual aetiology on the somewhat double-edged grounds that war paralysis, phobia and obsession were suffered by many 'whose sexual life seems to be wholly normal and commonplace' (p. 913). War neuroses provided abundant evidence in favour of the validity of Freud's theory of forgetting, he insisted. Hypnotism risked producing directly just those states of dissociation which it should be the psychiatrist's most vital duty to avoid. 'Instead of advising repression and assisting it by drugs, suggestion or hypnotism, we should lead the patient resolutely to face the situation provided by his painful experience'. The patient's view 'should be talked over in all its bearings' (p. 914).

The war then provided a heightened sense of the theoretical and political failure of pre-existing conventions and models. According to Freud's close colleague, Sandor Ferenczi, psychiatry had foundered in the face of the war psyche, just as historical materialism had crumbled in the face of human nature, or more specifically the recalcitrance of Russian society after the Revolution. Ferenczi's paper in the 1921 psychoanalytic collection on the war neuroses began with discussion of the Russian Revolution and the consternation the Soviet leaders experienced at finding that the new social order could not be so easily imposed as they had hoped. 'Finally they agreed that perhaps the materialistic idea was after all too one-sided', and 'A somewhat similar thing has occurred among neurologists during the war'.[80] What was evident in both socialism and psychiatry, he suggested, was a failure of theory which psychoanalysis could now redress:

> The mass-experiment of the war has produced various severe neuroses, including those in which there could be no question of a mechanical influence, and the neurologists have likewise been forced to admit that something was missing in their calculations – and this something was again – the psyche. (p. 6)

After examining various organic ideas on shellshock, speculation for instance about concussion, sudden pressure of the cerebro-spinal fluid or compression of the spinal cord, Ferenczi turned to psychoanalytic explanations. Shellshock, it seemed to him, stemmed from the damming up and the displacement of the libido.

Freud and Ferenczi both argued that traumatic neuroses such as shellshock demonstrated with a particular clarity the wider truths of psychoanalysis: 'The closest analogy to this behaviour of our neurotics is afforded by illnesses which are being produced with

80   Ferenczi *et al.*, *Psychoanalysis and the War Neuroses*, p. 5.

special frequency precisely at the present time by the war – what are described as traumatic neuroses.'[81] Freud differentiated between 'spontaneous neuroses' in general and traumatic neuroses, such as shellshock or other shock-induced states, for instance those caused by railway accidents. But despite the differences, all such phenomena had certain common features:

> But in one respect we may insist that there is a complete agreement between them. The traumatic neuroses give a clear indication that a fixation to the moment of the traumatic accident lies at their root. These patients regularly repeat the traumatic situation in their dreams; where hysteriform attacks occur that admit of an analysis, we find that the attack corresponds to a complete transplanting of the patient into the traumatic situation. It is as though these patients had not finished with the traumatic situation, as though they were still faced by it as an immediate task which has not been dealt with; and we take this view quite seriously. It shows us the way to what we may call an *economic* view of mental processes. Indeed, the term 'traumatic' has no other sense than an economic one. We apply it to an experience which within a short period of time presents the mind with an increase of stimulus too powerful to be dealt with or worked off in the normal way, and this must result in permanent disturbances of the manner in which the energy operates.[82]

Ernest Jones, founder of the British Society of Psychoanalysis in 1919, had campaigned throughout the war for a wider recognition of Freudian insights into trauma, mental illness and the causes of war in general. Writing in the *Sociological Review* in 1915, he pondered whether the science of psychology can ever show us how to abolish war. Jones introduces the work of Freud in this context and suggests that notions of the unconscious, the repressed and the inevitable conflict of civilisation and the instincts should be brought to bear in the treatment of the war victims: 'All the emotions of which we become aware, either in ourselves or in others, represent only tricklings through from the volcanic reservoir that is pent up in the unconscious region of the mind.'[83] The shellshocked soldier

81 Freud, *Introductory Lectures on Psycho-Analysis*, vol. XVI, p. 274.
82 *Ibid.*, pp. 274–5; *cf.* pp. 381–2.
    The use of painful electric shock treatment for shellshock victims, Freud observed in his 'Memorandum on the electrical treatment of war neurotics', ostensibly succeeded in causing the patient to flee back into health. Such a procedure of shocks appeared brilliantly successful at first, but proved disappointingly transient: the patient tended to relapse.
83 Jones, 'War and individual psychology', p. 169.

suffered from an explosive crisis which was matched by the volcanic nature of history itself. Beneath the conventional accounts of war lay deeper forces, 'some set of recurrently acting agents, which tends to bring about war more or less regularly, and to find or create pretexts for war whatever the external situation may be' (p. 170).

Jones speculates on whether there may not be an inevitable historical oscillation between war and peace, the latter being nothing but a temporary recoil (enduring perhaps a couple of generations) from the warlike impulses. It is widely acknowledged, he points out, that war constitutes a relapse from the standards of civilisation: 'Civilised warfare is a contradiction in terms, for under no circumstances is it a civilised act to blow another person's head off or to jab a bayonet into him' (p. 176). No nation today is able to admit its own war spirit or brazenly to declare itself the instigator of war: the other side is always taken to be responsible. War is justified as an attempt to restore peace; it is fought in the name of its own abolition: wars to end war. But, as Jones observes, the 'end justifies the means' argument masks the complex emotional investment in the *means*: 'we are led to ask whether the terrible events of war, the cruelties and so forth, are not connected with the underlying causes of war itself' (p. 175).

Even the 'good' motives for war, such as patriotism, involve complex emotional displacements: 'The ultimate psychological origin of this complex sentiment [patriotism] is to be found mainly in the individual's relation to his parents' (*ibid.*). Repressed instincts play a cardinal part in war. There is a widespread passion for cruelty, destruction, lust and loot. The motives for enlisting are equally if not still more complex and various, 'from the fascinating attraction of horrors to the homosexual desire to be in close relation with masses of men' (p. 177). Since the unconscious cannot be abolished, society needs to find suitable 'outlets' for the primitive emotions. Jones reiterates here an argument which as we saw in the introduction had already been expressed by William James in his essay, *The Moral Equivalent of War*. Jones seemed unsure that an 'equivalent' was currently possible, confessing that 'It is at present quite an open question whether it is possible for mankind to abstain from war' (p. 179).

Jones's article largely avoids any simple patriotic identification of its own. It impressively refuses to rejoin the standard arguments or platitudes either for or against war. But as though mirroring the subject matter – relapse – the text finally veers back into an apparent idealisation of military conflict, endorsing the notion of the grandeur of the contemporary crisis in Europe. Against the grain of his own argument, he concludes puzzlingly with a lame, 'even-handed' description of the heights and depths of war, reproducing something of the cliché of war as 'deeper' and more 'real' than peace, skimming over the complexity of what that supposed heightening of 'reality' involves and how it is achieved.

> [War] reveals all the latent potentialities of man, and carries humanity to the uttermost confines of the attainable, to the loftiest heights as well as to the lowest depths. It brings man a little closer to the *realities* of existence, destroying shams and remoulding values. (p. 180)

SEISMIC SHIFTS?

That idea of war as 'revelation', 'life', even 'purification', so elaborately theorised in the nineteenth century, found wide expression in culture of the First World War period: military conflict cast as vibrant, vital, natural battle, quickening the life force of the nation.[84] The terms were frequently Darwinian and the recurrent conclusion was war's ineluctable nature. It was this vision of power, the historian Mommsen observes, which led Max Weber, half against his will, to admire the Prussian junker class and especially its chief representative, Bismarck.[85] As Weber was to say in a lecture delivered in Munich on 22 October 1916, Germany was fighting a war for honour, rather than for mere gain or expansion. The war was the realisation of Germany's world-historical destiny, a form of 'consecration'.[86]

It is this evocation of war as consecration and truth, with its disturbing echoes, which I have tried to highlight. The historical representation of the First World War often unwittingly inherits something of that model, perpetuating the myth of war as a decisive one-way passage in the psychological 'life' of a culture or nation. Whilst I do not dispute that wars — least of all perhaps this century's world wars — have involved new and terrible historical, cultural and

---

84 To take an extreme but not untypical example, consider an essay in *Die Neue Rundschau*, which put it like this in November 1914: 'War! It was purification, liberation that we experienced, and an immense hope. It was of this that the poets spoke, only of this. What do they care for imperial power, for commercial hegemony, even for victory? ... What inspired the poets was the war in itself, as visitation, as ethical exigency' (quoted in Timms, *Karl Kraus*, p. 306).
85 See Mommsen, *Max Weber and German Politics*, pp. 38–40.
86 Quoted in Timms, *Karl Kraus*, p. 307.

social experiences, singularising evolutionary and evaluative schema are too often superimposed upon those complex phenomena. The First World War is often cast as a kind of sublime if terrible realisation – a vortex, threshold, royal road, origin, deluge, or midwife to history; a stark sweeping away of myths and self-delusions, an inevitable destiny to which the nineteenth century was leading, but of which it supposedly knew nothing.

I have sought to suggest that we hesitate, pondering the applicability of such conventional metaphors and ideas of war as destiny or as freewheeling machine. These notions, whilst often compelling, may well be obfuscating rather than helpfully clarifying, begging too many questions, borrowing too easily and fatalistically from the languages of the past. Is it so self-evident that, as W.B. Gallie, former Professor of Political Science at Cambridge University, puts it, war in general is an 'inherently cumulative process', possessed of an 'escalating dynamism' which has world war as its historical conclusion?[87] Gallie proposes a teleological reading of history culminating in the 'inevitability' of the twentieth century, whose essence is now understood as war: 'the whole thrust of our war culture across the ages has been *towards* this apogee' (p. 66).

Consider a very different example of this tendency. The French theorists Deleuze and Guattari describe the dynamic of war in terms of the expansive tendency of capitalism. Their account in this respect is arresting, yet also glib. War, we are told in very general terms, seeks to expand to its own limit, to widen its field as far as possible, like an inexorably advancing machine. The war machine is understood in this account as anterior to the state. The state possesses it and is possessed by it. Yet it belongs both anthropologically and conceptually to a nomadic world. Although the state appropriates the war machine, it can never quite subordinate it: 'As for the war machine in itself, it seems to be irreducible to the State apparatus, to be outside its sovereignty and prior to its law: it comes from elsewhere. . . . *The State has no war machine of its own*; it can only appropriate one in the form of a military institution, one that will always cause it problems.'[88] The war machine is always possessed of its own intrinsic energy and momentum:

> The problem is that the exteriority of the war machine in relation to the State apparatus is everywhere apparent, but remains difficult to conceptualize. It is not enough to affirm that the war machine is external to the apparatus. It is necessary to reach the

87   Gallie, *Understanding War*, pp. 51–3.
88   Deleuze and Guattari, *Nomadology: The War Machine*, p. 2.

point of conceiving the war machine as itself a pure form of exteriority, whereas the State apparatus constitutes the form of interiority we habitually take as a model, or according to which we are in the habit of thinking. (p. 5)

Historically, Deleuze and Guattari insist, the state seeks to tame and discipline the war machine but that does not mean it owns it or originates it. The war machine always retains something nomadic, something of 'the tribe in the desert': 'And each time there is an operation against the State – subordination, rioting, guerrilla warfare or revolution as act – it can be said that a war machine has revived, that a new nomadic potential has appeared' (p. 60).

Both the guerrilla attack on the state and the war machine of the state possess the same wandering, excessive quality: 'this new nomadism accompanies a worldwide war machine the organization of which exceeds the State apparatuses, and passes into energy, military-industrial and multinational complexes' (p. 62). For Deleuze and Guattari, as also for Foucault, war is not politics enacted by other means, but vice versa. Their critique suggests that the Clausewitzean relationship between war and politics needs to be turned upside down. War is the subject, politics merely one manifestation, one form war may adopt. As Foucault writes, 'power is war, a war continued by other means'.[89] The war machine circumscribes both the state of war and peace; indeed it does away with the sharp division between them. The state is now cast as the instrument of the war machine:

> This is the point at which Clausewitz's formula is effectively reversed; to be entitled to say that politics is the continuation of war by other means, it is not enough to invert the order of the words as if they could be spoken in either direction; it is necessary to follow the real movement at the conclusion of which the States, having appropriated a war machine, and having adapted it to their aims, reissue a war machine that takes charge of the aim, appropriates the States and assumes increasingly wider political functions.[90]

In an interesting essay considering the post-1945 nuclear world, Jacques Derrida is suspicious of the tendency to make links to previous discourses and material conditions of weaponry. He raises the question of whether the discovery of such continuities serves to anaesthetise us to the novelty of what has been 'invented' since, to

89 Foucault, *Power/Knowledge*, p. 90.
90 Deleuze and Guattari, *Nomadology: The War Machine*, p. 119.

recuperate the strangeness of the nuclear age to the familiarity of earlier models. On the one hand it can be argued that there is perhaps no invention, no radically new predicate in the situation known as 'the nuclear age'. It can be claimed that we are witnessing only 'the brutal acceleration of a movement that has always already been at work'. On the other hand, 'the critical zeal that leads us to recognize precedents, continuities, and repetitions at every turn can make us look like suicidal sleep-walkers, blind and deaf alongside the unheard-of'.[91]

But if the presumption of continuities in the historical discourse of the war machine arouses suspicion, so too, as Derrida makes clear elsewhere, does the idea of absolute divisions:

> I do not believe in decisive ruptures, in an unequivocal 'epistemological break', as it is called today. Breaks are always, and fatally, reinscribed in an old cloth that must continually, interminably be undone. This interminability is not an accident or contingency; it is essential, systematic, and theoretical. And this in no way minimizes the necessity and relative importance of certain breaks, of the appearance and definition of new structures.[92]

Derrida counterposes the hardware and the discourse, the technical and the discursive aspects of weaponry; he insists on the need to distinguish the extraordinary new sophistication of nuclear technologies from the old sophistry, rhetoric and vulgar psychology which informs and surrounds them: 'Between the Trojan war and nuclear war, technical preparation has progressed prodigiously, but the psychagogic and discursive schemas, the mental structures of intersubjective calculus in game theory have not budged.'[93] The recuperation of what is new in the nuclear situation to the pre-nuclear context, he suggests, may easily become a kind of defence against or domestication of the anxieties appropriately commensurate with new realities – the new reality of the potential destruction of history itself – of the very archive as such. Recuperation and assimilation (for example of the apocalyptic or bimillenarist type) might make us blind and deaf to the newness of what we face. One needs to cut through the apparent resemblance of this discourse to all others.

The opposition set up in Derrida's essay between the technical and the discursive dimensions of the nuclear is salutary, but must also be seen as tenuous. As he himself points out, the whole tech-

91 Derrida, 'NO APOCALYPSE, NOT NOW', p. 21.
92 Derrida, *Positions*, p. 24.
93 Derrida, 'NO APOCALYPSE, NOT NOW', p. 24.

nical apparatus is to an unprecedented degree textual, *'fabulously textual*, through and through'. Nuclear weaponry depends, more than any weaponry in the past, it seems, upon structures of information and communication, language, including non-vocalisable language, codes and graphic decoding. But the phenomenon is 'fabulously textual' also to the extent that beyond Japan 1945, a nuclear war has not taken place: one can only conceive of it, hypothesise about it in its absence, use the images and narratives we know.

Total nuclear war is an imaginary figure – signifier of the unthinkable signified. To talk or think of the total nuclear catastrophe is in a sense to talk of the impossible. To speak of the final nuclear disaster is to give the entirely other a more or less familiar meaning, to bring it within the intertextual, to evoke the traces, for example, of a past eschatological literature. As a result, 'The growing multiplication of the discourse – indeed, of the literature – on this subject may constitute a process of fearful domestication, the anticipatory assimilation of the unanticipable entirely – other' (p. 23). There is a constant circulation between the structuring narratives of history and the absolutely concrete fact of the weaponry – its mass destructive potential: 'For the "reality" of the nuclear age and the fable of nuclear war are perhaps distinct, but they are not two separate things. It is the war (in other words the fable) that triggers the fabulous war effort, the senseless capitalization of sophisticated weaponry, this speed race in search of speed . . .'. This 'crazy precipitation', Derrida suggests, 'structures not only the army, diplomacy, politics, but the whole of the human *socius* today' (*ibid.*).

The perception of society today structured as a psychotic war machine and the insistence on 1945 as the key break has other proponents. Exploring the continuities and differences between war in general and nuclear war in particular, the Italian psychoanalyst Franco Fornari explores the concept of psychological defences against reality. Whilst the nuclear machine is at one level the most massive defence against danger, paradoxically it begins to break down the function of war as psychic defence. Borrowing from Karl Kraus, one might say it is the fatal symptom of the disease it purports to cure. Or in Fornari's terms, it is the very carrier of the psychotic force against which it seeks to protect its citizens. His argument assumes that war is 'primarily the result of psychotic processes'.[94] This is a view which has long been supported by

94   Fornari, *The Psychoanalysis of War*, p. 7.

other psychoanalytic commentators. Roger Money-Kyrle went still
further when he observed before the Second World War:

> The psychotic mechanisms of infancy, therefore, are very far
> reaching in their effects. They are among the necessary conditions
> not only of superstition, but also of social solidarity, of warfare,
> and even of the arts and sciences. To them we largely owe both
> what we admire and what we regret in our culture. . . . They are
> among the most fundamental characters of the human species;
> more fundamental even than intelligence. . . . Therefore the main
> difference between man and other animals is not so much his
> greater intelligence, but the defects peculiar to his intelligence,
> almost one may say his capacity for madness.[95]

In the twentieth century, Fornari argues, war becomes increasingly
less capable of controlling deep psychotic anxieties (p. xxii).
Fornari's work offers perhaps the most sustained post-Freudian
psychoanalytic engagement with war, although there is a whole
range of significant literature on war spanning the entire period
from the 1930s onwards which I cannot encompass here.[96] The
recent work of Hanna Segal on nuclear war also warrants serious
consideration. Like Fornari she stresses the psychotic dynamics of
so much 'defence talk' and, in a memorable observation highly
pertinent to the central 'machine' theme of my study, she questions
the terms of our technological determinism. It is worth quoting the
passage in its entirety:

> The growth of technology is also used for a typically schizoid
> dehumanization and mechanization. There is a kind of pre-
> vailing depersonalization and de-realization. Pushing a button to
> annihilate parts of the world we have never seen is a mechanized
> split off activity. Bracken [in *The Command and Control of
> Nuclear Forces*, 1984] contends that the war is likely to happen
> through our machines getting out of control. Everything is so
> automated that over-sensitive machines could start an unstop-
> pable nuclear exchange. The MIT computer expert, Joseph

95 Money-Kyrle, *Superstition and Society*, p. 114.
96 From the 1930s, one might cite, for instance, Glover's *War, Sadism and
    Pacifism* and (with Ginsberg) his 'Symposium on the psychology of peace and
    war', as well as Money-Kyrle's wide-ranging BBC broadcasts and essays, such
    as, 'A psychological analysis of the causes of war' and 'The development of
    war. A psychological approach', both republished in his *Collected Papers*.
    Money-Kyrle's discussion ranges from the self-sacrifice of war to shellshock,
    from infantile fantasies and projections to the concept of the insanity of
    nations themselves in wartime. Mention should also be made of Alix
    Strachey's *The Unconscious Motives of War* (1957).

Weizenbaum, comes to the similar conclusion: that modern big computers are so complicated that no expert can see through and control them. The whole nuclear early warning system is based on these machines – perhaps the worst danger if the paranoid international tensions reach a high level. Since one effect of nuclear explosion is a disturbance in communication systems, it might not be in the powers of governments to stop an escalation even if we wish to. But the fact that we can even think it – 'machines will start the war, not us' – shows the extent of the denial of our responsibility. We seem to live in a peculiar combination of helplessness and terror and omnipotence – the helplessness and omnipotence increasing one another in a vicious circle. This helplessness which lies at the root of our apathy is partly inevitable. We are faced with a horrifyingly threatening danger. But partly it is self-induced and becomes a self-fulfilling prophecy. Confronted with the terror of the powers of destructiveness, we divest ourselves from our responsibilities by denial, projection and fragmentation.[97]

Segal explores the vicious circles of hatred and fear which operate in nuclear deterrence theory and likens them to such phenomena in a clinical setting. Groups may also have their own psychotic features. The collective constitution of the enemy as sub-human and the disavowal of the real dangers of the arms race itself, she suggests, are analogous to an individual regression to the 'paranoid/ schizoid' position, characterised by the operation of denial, splitting and projection. Just because nuclear war so decisively threatens the belief in a symbolic survival after death, it has to be denied, split off altogether. Truly to face the idea of all-out nuclear conflict is to abandon the conventional compensations of death, the very possibility of an afterlife of meaning and memory in another's mind. Nuclear war mobilises the wish for and the utter dread of annihilation. 'Those of us who work with psychotics get an inkling of this kind of terror' (p. 7). The language of nuclear war (for example the personalisation of the bomb dropped on Nagasaki as 'Fat Man', or the abbreviation 'Nuke', to say nothing of the word 'deterrence' itself) serves to domesticate the unspeakable terror.

Segal is keen to emphasise the specific existential situation and the particular psychological corollaries of the nuclear context. Characteristically, however, psychoanalytic writing tends to universalise the function of war in a manner which does not adequately discriminate between particular (and arguably quite unrelated) his-

97 'Silence is the real crime', pp. 9–10.

torical forms of conflict. Even more contentiously, the history of war across the ages is sometimes taken to match the psychic history of the individual in the Freudian scenario. Ontogeny recapitulates phylogeny for Freud, whilst the reverse is true for Fornari in the 'psychotic' history of war.[98] Compare the historian Tim Blanning, who seeks to shift the locus of the universal from war to violence: 'It is violence which is universal, not war, which is just one very specialised form of violence.'[99] Blanning points out that if 'human nature' was the cause of war in 1914, it was the cause of peace in 1910 – and whatever else changed in those four years it was not human nature. He rejects the idea that war is caused by or simply 'reflects' some deep atavistic popular pressure.[100] For Fornari, however, the key question is not who causes war, but what *psychic function* war has. War is a social manifestation of an individual structure of mind, he suggests. It 'represents a social institution the aim of which is to cure the paranoid and depressive anxieties existing (in a more or less marked degree and more or less resolved in terms of integration with reality) in every man'.[101] War is a security system, a defence against internal terrors and anxieties. That function, paradoxically, is 'not to defend ourselves from an external enemy, but to *find a real enemy*' (p. xvii).

Whatever the universal or contingent attributes of war, I have sought to provide here a certain historicisation of such an argu-

---

98   Two stages of explanation are sharply contrasted by Fornari, the universal and the contingent; psychoanalysis then situating itself firmly on the side of the former. Fornari declares firmly that psychoanalysis is metahistorical and metapolitical (*The Psychoanalysis of War*, p. xxvi).

Where history does come in, it tends to be viewed structurally. We are offered a theory of *stages*, and the reiteration of Freud's contention that there is an increasing discrepancy between 'progress' and 'instinct', civilisation and its discontents. Fornari observes *in history*, a disconcerting regressive psychic evolution matched by increasingly sophisticated technical advances in armaments. Thus primitive warfare (lances, swords) can be understood symbolically as the desire for insertion into the body of the other (genital-sadism), the advent of firearms constitutes the realisation of fantasies of projectiles ejected towards the enemy (anal sadism), whilst the chemical and the nuclear realise fantasies of omnipotence and annihilation (oral sadism).

99   Blanning, *The Origins of the French Revolutionary Wars*, p. 4.

100 According to Blanning: 'it is very difficult to find *any* war which was launched as a result of popular pressure. It played a certain part in the War of Jenkins' Ear, the American War of Independence, the French Revolutionary War of 1792, the Crimean War, the Franco-Prussian War, perhaps even the First World War, but however generous one's category, the list would not be long' (*The Origins of the French Revolutionary Wars*, p. 5). On the lack of evidence for French popular pressure as a cause of war in 1914, see Becker, *1914*.

101 Fornari, *The Psychoanalysis of War*, pp. xv–xvi.

ment; to make a link between this view today and powerful currents of thought in the century or so before 1939. But psychoanalysis cannot rightly be viewed as the mere recapitulation of earlier thought on war. The examples I have sketched from Freud, Riviere, Fornari and others *do* offer new terms of understanding, but their work is clearly part of a much broader network of European thought on conflict. The language of war in the nineteenth century frequently focused on war's symbolic function in the construction of identities, borders and boundaries; war's role both in mobilising and displacing anxieties, 'unrest' and so forth. Beneath the surface rationales of war, Fornari reminds us, the real aim is to externalise an enemy, through a manoeuvre *'which transforms this terrifying but ultimately* unaffrontable entity into an external, flesh-and-blood adversary who can be faced and killed' (p. xvi). His approach raises many fruitful themes and questions about 'applied psychoanalysis', but also problems of historical location. As he himself points out: 'it may be affirmed that psychoanalysis can explain, at most, some of the mechanisms of war in general but not why war breaks out at a particular moment' (p. xxiv). Psychoanalysis, it is suggested, may be able to probe the discrepancy and the interaction between the politico-economic interests in war on the one hand and the realms of fantasy, anxiety, desire, psychotic process, paranoia and xenophobia on the other: 'psychoanalytic research could acquire a particular function in revealing the nature of the schizo-paranoid processes that lie beneath what can be called the group's general tendency to form in a Manichean way' (p. 253). In the nuclear age, he observes, questions about the 'real' material conflict and the psychic purpose, and about internal/external balances/imbalances of terror, are interfused perhaps more deeply and inextricably than ever before.

In both Derrida's and Fornari's account, war and the technology of violence can no longer be seen as a mere instrument of political power, economic interest or social control (although it may include all of those possibilities). The war machine is perceived to have its own momentum, dislocated from 'Reason', even economic interest, but not without motivation or causation. The idea of war as crazy anarchic machine is in one sense anathema to the Clausewitzean view with which I began, in which war should apparently never be thought of as something autonomous but always as an instrument of policy.[102] For Clausewitz war is purposive, instrumental, reducible not only to positive laws of functioning, but also to a kind of ideal science of planning which includes anticipating the friction

102 See Clausewitz, *On War*, Bk. I, ch. 1.

and irrationality of participants in the heat of battle. For Clausewitz, war, for all its friction, is amenable to laws and patterns. It remains functional for the state and amenable to a kind of 'natural scientific' exploration.

Between 1815 and 1918 we can find numerous other endeavours to produce a definitive 'in-depth' science of politics and international conflict. When the biologist and social scientist Patrick Geddes addressed the issue of science and war in the *Sociological Review* in 1915, for example, he recognised the difficulty of the task although he quickly took refuge in a flurry of labels: 'Why not then a science of war – a branch of sociology of course, but still a distinctive and rational strategology; or, more exactly, a description of war or strateography, with an interpretation of their significance – as far as may be indeed, a strategosophy.'[103] As Geddes goes on to admit: 'At first, no doubt the problem seems insuperably difficult. It is the problem of describing a great game of life and death for multitudes, in fitful labyrinthine progress since man, and indeed his progenitors, have been upon the earth' (p. 16). Yet, despite those apparent difficulties, Geddes is still hopeful that there must be a certain historical cycle: certain patterns of periodicity: 'in the tangled and seemingly confused chronology of history are there not observable some periodicities of recurrence, waves corresponding to generations and semi-generations?' In short, 'May not the history of wars be more orderly than it has seemed?' (*ibid.*)

Geddes introduces a whole set of techno-scientific terms (which I readily confess I have not fully fathomed) to characterise the relations of order and disorder; war is said to be 'geoklastic', 'paleotechnic' and 'kakopsychic' whilst peace is 'geotechnic' or 'Eupsychic' (p. 22). He suggests that a science of war might correspond to an earth science capable of detecting impending eruptions – 'an orderly vulcanology and seismology' (p. 16). He seeks to separate out the fundamental from the trivial battle; to isolate the truly 'nodal wars' from the mere surface movements. The problem of war, like the problem of volcanoes, Geddes acknowledges, is that things may be going on under the surface, beneath the apparently stable crust, during periods of relative calm: war and peace cannot be simply counterposed; peace may be latent war. 'Our commonly so-called "peace" [may be] mere war static and potential'. States of peace may be nothing but 'rhythmic phases of what we should call Wardom' (p. 21). To deny war, Geddes seems to hint, is to forget the subterranean war world of psychic life:

103   Geddes, 'Wardom and peacedom', p. 15.

those who believe in pacifism have to answer the charge that 'it is in Wardom that we live, move and have our being' (*ibid*.).

The dream of an objective science of war persists. Thus the French social scientist and demographic commentator of the 1930s, Gaston Bouthoul, invents the field of 'Polemology' after 1945. For all its good intentions and sometimes its interesting observations, this often seems to be a strangely naive constellation of war generalisations, demography, statistical methods and structuralist ambitions. There are of course other French projects one might contrast – more influential than Bouthoul's approach has been Raymond Aron's exploration of the relative openness of frontiers and intellectual exchanges in different periods, and the conduciveness of different international systems to war and peace. I focus on Bouthoul, however, because his work seems to rejoin in so flagrant a way an earlier aspiration of social science. He hopes to produce both a scientific taxonomy and a profound chronology of war. The plethora of causes is to be refined into accessible generalisations. He strives for precise divisions and tabulations, sorting out war's periodicity, intensity, typologies; distinguishing ultraconflicts, macroconflicts, microconflicts, infraconflicts. Polemology, we are told, 'seeks to analyse and interpret the structural causes (demo-economic, geographic, mental . . .) which engender collective aggressivity'.[104] The interaction of the social and the psychological are to be discovered in the seismic shifts and continuities of 'mentalities' over the long term. The project is at once synchronic and diachronic: tracing out the structural features of war at a given time across the globe and periodising across history the key structuring moments – the decisive turns in the passage of war.

Bouthoul suggests three grand mutations in the history of war, 1775, 1914 and 1945. In the American War of Independence (1775–83), an extra-European colony for the first time revolted successfully *tout court* against the European metropolis; it opened up the idea of national self-determination, a people's army, a *levée en masse*. Wars began to interconnect ever more powerfully with total economies, a process which culminated in 1914–18 and the effective symbiosis of the economic and the military. 1914, it is argued, constitutes the second key moment effecting a fundamental economic and political restructuring of the world, announcing the decline of Europe and the rise of the Third World – hence a 'League of Nations' was to replace 'The Concert of Europe'. 1945 was the third moment of transformation for three reasons: nuclear power, demographic crisis, and new forms of speed. The last

104   Bouthoul and Carrere, *Le Défi de la guerre*, p. 34.

of these factors is to be understood both materially and psychologically: an incredible escalation of the idea and the material reality
of speed apparently separates the pre- and post-1945 world.

Under the double shock of the Second World War and the
nuclear explosions of 1945, Bouthoul established a new institutional apparatus for the study of international conflict under the
aegis of science. But far from addressing the personal sense of
disturbance and urgency which prompted the project, Bouthoul
would abandon the disagreeable unquatifiability of subjective feeling
altogether and turn to tables, statistics, 'science', neutral, cool
observation. Whilst it would be an exaggeration and unfair to
remark, as Keegan remarks of the official history of the First World
War, that it 'achieved the remarkable feat of writing an exhaustive
account of one of the world's greatest tragedies without the display
of any emotion at all',[105] nevertheless this 'polemology' also
represses any sense of internal conflict from its own account. The
project endeavours to achieve complete political and emotional
detachment, the avoidance of 'politicisation and polemic, both
of which are contrary to the scientific character (logos) which
[polemology] seeks to preserve'.[106] Polemology 'thus presents itself
as the scientific study of war' (p. 34).

In an extreme way, Bouthoul's strategy exemplifies a problem in
all 'academic' discussion of war, including the present inquiry. It is
not clear what it would be to write of war *as an idea* or as a
language without the very activity constituting a kind of intellectual
distancing and denial. Blood and guts, shells, trenches and radiation
are signifiers, but they cannot be considered just in those terms lest
we return, grotesquely, to a vision of war as some mere chessboard.
Discourse is a part of the (but not the only) materiality of war. One
must insist on the cultural encoding of war, the fact that there is no
simple 'unmediated' experience, no pure 'bellicose instincts', but
one must also recognise that the very fact of such analysing and
verbalising may constitute a kind of neutralisation and domestication. Derrida suggests that to write or 'understand' total nuclear
war may mean, by definition, to defend oneself against a future
which cannot be experienced or thought ahead of its own occur-
rence. If the representation of the nuclear poses special difficulties, it
is also part of a wider problem of incommensurability, a wider
ethical dilemma about intellectual or scientific investigations of
war – the risk, or perhaps the inevitability, as here too in my

105   Keegan, *The Face of Battle*, p. 29.
106   Bouthoul and Carrere, *Le Défi de la guerre*, p. 34.

discussion, of detaching the actuality of wars – what wars do to people – in the circumlocutions of writing and critique.

For Bouthoul the distancing is the aim – the detachment of 'science'. Yet his text is unable – whether willing or not – to avoid politicisation, polemic or indeed the disclosure of intense anxiety. Its periodisation of war history stresses 1945 not only because of Hiroshima and Nagasaki but because in Bouthoul's view that year was the moment when the Third World declared war on the West with the demographic 'time bomb'. The confusion and the disturbance of the war are evacuated into a nightmare of remote, foreign bodies exploding over the globe, encroaching on the West.[107] Whatever the urgent realities of demographic crisis, to slide so easily and compulsively from war to Third World population 'bomb' cannot go unremarked – it joins a long Malthusian tradition.

In Clausewitz, the desire for a positive science of war resided alongside an announcement of the passing of the historical conditions and the material basis for any such over-rationalistic representation. Yet Clausewitz was sometimes to evoke war as a kind of princely card game – a raising of the stakes of politics, a calculated gamble, exactly at the historical moment when, so it would seem, the social reality of war in Europe had overrun any such image. A century later, the First World War undermined for some although not all theorists (think of Lenin) a belief in the purposiveness of war as Clausewitz had formulated it – war as the furtherance of taken-for-granted interests. But that in turn rejoined another element of the Clausewitzean script of war: irreducible chaos; the explosion of fixed laws, the crisis of crisis management. Clausewitz, for all his protestations of the subservience of war to politics, provides a shadow story in which war destroys politics.

What I have tried to suggest is that to read the nineteenth-century war writings of, amongst others, Clausewitz, De Quincey, Proudhon and Ruskin is to read a complex set of discourses about power and anarchy, the deep functions of war, the interrelationship of industrialisation, modernity and destruction, the freewheeling nature of machinery, the forces that could be marshalled for and against disintegration. Moreover this nineteenth-century literature was caught up in a complex dilemma about method – about how to write adequately of war, how to think its terms, how to engage with it deeply. Clausewitz acknowledged the operation of 'moral' factors in war (if not the moral issue of war itself), referred to the original violence and animosity of its impetus, the huge difference between mathematics and medicine, let alone the sphere of the 'physician of

107 See *ibid.*, p. 88.

the mind', but then moved the discussion on, having neither the language nor the inclination to engage those issues as his central concern.

In that context then, Einstein's gesture in turning to Freud in 1932 was not without its historical symbolism. Clausewitz had insisted in 1832 that war could never be a geometry as the eighteenth century had once hoped; one hundred years later, Einstein would make the point more personally. Einstein was more than just a physicist – he was an icon. His brain, as Barthes so deftly suggests in *Mythologies*, has become a cliché of scientific genius. Einstein to some extent colluded in the myth by leaving his brain to posterity, or rather to two squabbling hospitals who wanted to do the definitive post-mortem on his genius.[108] In any event, invited to write a public letter to the correspondent of his choice and faced by the memory of the First World War and the lengthening shadows of Nazism, Einstein publicly turned to Freud, as though to say: natural science alone cannot solve human problems. Or to put it another way, physics wrote a letter to psychoanalysis, imploring an answer by return of post to the question, 'why war?' Perhaps Freud, Einstein seemed to hope, might get to grips with the bewildering friction that even Clausewitz had admitted was incalculable.

108 See Barthes, 'The brain of Einstein', in *Mythologies*.

Bibliography

Place of publication is London unless otherwise stated.

PAMPHLETS

Angell, Norman, *The Prussian in our Midst*, 1915.
Anon., *The Channel Tunnel. A True View of It! Regarded as a Great Whole. (For Popular Verse)*, 1882.
Anon., *The Channel Tunnel by a Military Railway Expert*, 1907.
Anon., *What Caused the War. German Policy Explained by a German*, issued by the Central Committee for National Patriotic Organisations, n.d.
Anon., *Shall We Go on? A Socialist's Answer*, 1918.
Archer, William, *Fighting a Philosophy*, Oxford, 1914.
Asquith, H.H., *A Call to Arms*, a speech by the Prime Minister at the Guildhall, 4 September 1914.
—— *The War of Civilization*, a speech by the Prime Minister in Edinburgh, 18 September 1914.
Barker, Ernest, *Nietzsche and Treitschke. The Worship of Power in Modern Germany*, Oxford, 1914.
—— *The Submerged Nationalities of the German Empire*, Oxford, 1915.
—— *Linguistic Oppression in the German Empire*, 1918.
Bennett, W., *England's Mission*, Oxford, 1914.
Bradlaugh, Charles, *The Channel Tunnel: Ought the Democracy to Oppose or Support It?*, 1887.
Brailsford, H.N., *The Origins of the Great War*, Union of Democratic Control, n.d.
Chamberlain, Houston Stewart, *Who is to Blame for the War?*, New York, 1915.
'Chauvin, Henry' [pseud.], *The Battle of Boulogne or How Calais Became English Again*, n.d.
Collins, William J., *The Semeiology of the World-Wide War*, reprinted from *Scientia*, XXII (1918), Bologna, 1918.
Conrad, Joseph, *The Shock of War through Germany to Cracow*, printed for private circulation, 1919.
—— *Autocracy and War*, printed for private circulation, 1919.
—— *The North Sea on the Eve of War*, printed for private circulation, 1919.
Cromer, The Earl of, *Pan-Germanism*, reprinted from *The Spectator* (25 September 1915), 1916.
Cronau, Rudolf, *England as a Destroyer of Nations*, New York, 1915.
D'Arsac, *War and Peace: The German Snare*, no place or date given [1918?].
Dudon, Paul, *La Guerre: qui l'a voulue?*, Paris, 1915.
Durkheim, E. and Denis, E., *Who Wanted War? The Origins of the War According to Diplomatic Documents*, Paris, 1915.
Ford, Patrick, *The Criminal History of the British Empire*, no place of publication given, 1915.
Forth, C., *The Surprise of the Channel Tunnel. A Sensational Story of the Future*, Liverpool, 1883.

Francke, Kuno, *Germany's Fateful Hour*, Chicago, n.d. [1914?].

Frobenius, H., *Germany's Hour of Destiny*, issued by *The Fatherland*, New York, 1914.

Gladstone, W.E., *Channel Tunnel. Great Speech*, 1888.

Gooch, G.P., *The Races of Austria-Hungary*, 1917.

Gordon, Hanford L., *The Unholy War on Germany*, no place of publication given, 1915.

Gorham, Charles T., *The World War: Who is to Blame? A Reply to Professor Haeckel and Dr Paul Carus*, 1915.

Grey, Sir Edward, *The Peace of the World*. Speech delivered to the House of Commons, 13 March 1911.

'Grip' [pseud.], *How John Bull Lost London; or the Capture of the Channel Tunnel*, 4th edition, 1882.

Hanotaux, Gabriel *et al.*, *La Force brutale et la force morale*, preface by F. Laudet, Paris, 1915.

Harrison, Frederick, *The Meaning of War for Labour – Freedom – Country*, 1914.

Helfferich, Karl, *The Dual Alliance v.s. the Triple Entente. Germany's Case in the Supreme Court of Civilization*, New York, 1915.

Henderson, Yandell, *American Progressivism: Its relation to German Kultur from the Standpoint of a Progressive and Pro-German*, New York, n.d. [1916?].

Hobson, J.A., *The German Panic*, with an introduction by the Right. Hon. the Earl Loreburn, 1913.

James, William, *The Moral Equivalent of War*, American Association for International Conciliation, New York, 1910.

Kahn, Otto H., *Prussianized Germany. Americans of Foreign Descent and America's Cause*, no place of publication given, 1917.

—————— *The Poison Growth of Prussianism*, speech given in Milwaukee, Wisconsin 13 January 1918, no place of publication given.

Kipling, Rudyard, *Kipling's Message*, 1918.

Kropotkin, Pierre, *La Guerre*, Paris, 1912.

Lindsay, A.D., *War against War*, Oxford, 1914.

McClure, Canon, E., *Germany's War-Inspirers*, 1914.

Matheson, P.E., *National Ideals*, Oxford, 1915.

Merrifield, F., *Human Evolution in the Direction of Degeneration with Especial Reference to Prussianized Germany*, A Lecture to the Brighton and Hove Natural History and Philosophical Society, 12 January 1915, reprinted with some additions from the *Sussex Daily News*, 13 January 1915.

Morello, V., *L'Aggressione della Germania*, Rome, 1918.

Muir, Ramsay, *The Antipathy between Germany and England*, reprinted from *Scientia*, XVIII (1915), Bologna, 1915.

Muirhead, J.H., *German Philosophy and the War*, Oxford, 1914.

National War Aims Committee, *Aims and Efforts of the War. Britain's Case after Four Years*, 1918.

Pollock, Frederick, *German 'Truth' and European Facts about the War*, issued by the Central Committee for Patriotic Organisations, no place or date of publication given.

Rose, J. Holland, *Why we are at War*, 1914.

Russell, Bertrand, *War: The Offspring of Fear*, n.d.

Sadler, M.E., *Modern Germany and the Modern World*, 1914.

Sanday, W., *The Deeper Causes of the War*, Oxford, 1914.

Schrader, Frederick F., Introduction to *England on the Witness Stand. The Anglo-German Case Tried by a Jury of Englishmen*, New York, 1915.

Spielmann, Isidore, *Germany's Impending Doom. Another Open Letter to Herr Maximilian Harden, 4th August 1918, with four cartoons specially drawn by Hugh Thomson*, 1918.

Swanwick, H.M., *Women and War*, n.d.

—— *Frankenstein and his Monster. Aviation for World Service*, 1934.

Tittoni, Tommasso, *Who was Responsible for the War? The Verdict of History*, Paris, 1918.

Varinot, J., *La Paix et la guerre. Solution unique et rationnelle du problème*, Paris, 1905.

'Vindex' [pseudonym of G.W. Rusden], *England Crushed; the Secret of the Channel Tunnel Revealed*, 1882.

Vinogradoff, Paul, *Russia: The Psychology of a Nation*, Oxford, 1914.

Warren, Herbert, *Poetry and War*, Oxford, 1914.

Wilkinson, Spenser, *The Coming of the War*, Oxford, 1914.

ARTICLES

Anon., 'Foreign invaders', *All the Year Round*, NS 5 (1871), 133–8.

Anon., 'Sieges of London', *All the Year Round*, NS 5 (1871), 492–8.

Anon., 'Army organisation', *Macmillan's Magazine*, 23 (1870–1), 161–70.

Anon., 'Musings without method', *Blackwood's Magazine*, 194 (1913), 416–27.

Anon., 'Neurasthenia and shellshock', *The Lancet* (1916), 627–8.

Bath, Marquis of, *et al.*, 'The proposed Channel Tunnel. A protest', *The Nineteenth Century*, 2 (1882), 493–500.

Benison, Saul A., Barger, Clifford and Wolfe, Elin L. 'Walter B. Cannon and the mystery of shock: a study of Anglo-American co-operation in World War I', *Medical History*, 35 (1991), 217–49.

Blanning, Tim, 'Historiographical review. The death and transfiguration of Prussia', *Historical Journal*, 29 (1986), 433–59.

Bogacz, Ted, 'War neurosis and cultural change in England, 1914–22: the work of the War Office committee of enquiry into "shell-shock"', *Journal of Contemporary History*, 24 (1989), 227–56.

Boule, Marcellin, 'La Guerre', *L'Anthropologie*, 25 (1914), 575–80.

Brock, Arthur J., 'The re-education of the adult. I. The neurasthenic in war and peace', *Sociological Review*, 10 (1918), 25–40.

Burgess, John W., 'The present and future civilization of the world politically lies in the hands of the three great Teutonic states, Germany, England and the United States', *The Fatherland. A Weekly, Germany's Just Cause*, New York, n.d., pp. 5–7.

—— 'Reviewing the causes', *The Fatherland, Germany's Just Cause*, New York, n.d., pp. 23–6.

Campbell, Harry, 'The biological aspects of warfare', *The Lancet*, I (1917), 433–5, 469–71.

Carliol, J.W., 'The inner meaning of the war', *The Nineteenth Century and After*, 76 (1914), 730–6.

Chesney, George Tomkyns, 'The Battle of Dorking', *Blackwood's Magazine*, 109 (1871), 539–72.

Davin, Anna, 'Imperialism and motherhood', *History Workshop Journal*, 5 (1978), 9–65.

Dernberg, Bernard, 'Germany and the War. Not a defence but an explanation', *The Fatherland, Germany's Just Cause*, New York, n.d.

Derrida, Jacques, 'NO APOCALYPSE, NOT NOW (full speed ahead, seven missiles, seven missives)', *Diacritics* (Summer 1984), 20–31.

Dewar, Michael, 'A defence of Mutla Ridge', *The Guardian* (11 April 1991), 19.

Farquharson, John, 'After Sealion: A German Channel Tunnel?', *Journal of Contemporary History*, 25 (1990), 409–30.

The Fatherland. A Weekly, Germany's Just Cause, New York, n.d.

———— *Truth about Germany. Facts about the War*, New York, n.d. [1914?].

———— *The Viereck–Chesterton Debate on 'Whether the Cause of Germany or that of the Allied Powers is Just'*, New York, 1915.

———— *England on the Witness Stand. The Anglo-German Case Tried by a Jury of Englishmen*, New York, 1915.

Freud, Sigmund, 'Wir und der Tod', *Die Zeit*, 30 (20 July 1991), 42–43.

Gat, Azar, 'Clausewitz and the Marxists: yet another look', *Journal of Contemporary History*, 27 (1992), 363–82.

Geddes, Patrick, 'Wardom and peacedom: suggestions towards an interpretation', *Sociological Review*, 8 (1915), 15–25.

Glover E. and Ginsberg M., 'A symposium on the psychology of peace and war', *British Journal of Medical Psychiatry* (1933), 275–93.

Goadby, Edwin, 'A few words for Bismarck', *Macmillan's Magazine*, 23 (1870–1), 339–46.

Greenberg, Dolores, 'Energy, power and perceptions of social change in the early nineteenth century', *American Historical Review*, 95 (1990), 693–714.

Guiraudon, Virginie, 'Cosmopolitanism and national priority: attitudes towards foreigners in France between 1789 and 1794', *History of European Ideas*, 13 (1991), 591–604.

Hale, William Bayard, 'An American View', *The Fatherland, Germany's Just Cause*, New York, n.d.

Hart, R.C., 'A vindication of war', *The Nineteenth Century and After*, 70 (1911), 226–39.

Henderson, D.K., 'War psychoses: an analysis of 202 cases of mental disorder occurring in home troops', *Journal of Mental Science*, 64 (1918), 165–89.

Hoffman, Louise, 'War, revolution and psychoanalysis: Freudian thought begins to grapple with social reality', *Journal of the History of the Behavioural Sciences*, 17 (1981), 251–69.

Hopkins, Robert, 'De Quincey on war and the pastoral design of *The English Mail Coach*', *Studies in Romanticism*, 6 (1967), 129–51.

Irvine, Dallas D., 'The French discovery of Clausewitz and Napoleon', *Journal of the American Military Institute*, 4 (1940), 143–61.

Jackson, J.W., 'On the racial aspects of the Franco-Prussian War', *Journal of the Anthropological Institute of Great Britain and Ireland*, 1 (1871–2), 30–44, discussion, 44–52.

Jastrow, Morris, 'Appeal for fairness', *The Fatherland. A Weekly, Germany's Just Cause*, New York, n.d., pp. 12–14.

Jones, Ernest, 'War and individual psychology', *Sociological Review*, 8 (1915), 167–80.

Jones, Robert Armstrong, 'The psychology of fear and the effects of panic fear in war time', *Journal of Mental Science*, 63 (1917), 346–89.

Jordan, David Starr, 'War and manhood', *Eugenics Review*, 2 (1910–11), 53–60.

———— 'The eugenics of war', *Eugenics Review*, 5 (1913–14), 197–213.

Keay, John, 'The presidential address on the war and the burden of insanity', *Journal of Mental Science*, 64 (1918), 325–44.

Keith, Arthur, 'The Bronze Age invaders of Britain', Presidential address to Royal

Anthropological Institute, *Journal of the Royal Anthropological Institute of Great Britain and Ireland*, 45 (1915), 12–22.

Knowles, James, 'The revived Channel Tunnel project', *The Nineteenth Century and After*, 61 (1907), 173–5.

—— 'Imperial and national safety. (III) The revived Channel Tunnel project', *The Nineteenth Century and After*, 74 (1913), 469–71.

Lilla, Mark, 'The end of philosophy', *Times Literary Supplement*, (5 April 1991), 3–5.

Maudsley, Henry, 'War psychology: English and German', *Journal of Mental Science*, 65 (1919), 65–87.

Melville, C.H., 'Eugenics and military service', *Eugenics Review*, 2 (1910–11), 53–60.

Monod, Gabriel, 'Souvenirs of the campaign of the Loire', *Macmillan's Magazine*, 24 (1871), 69–80, 134–43.

Mott, F.W., 'Mental hygiene in shell-shock during and after the war', *Journal of Mental Science*, 63 (1917), 467–88.

Nicholls, David, 'Richard Cobden and the International Peace Congress Movement, 1848–1853', *Journal of British Studies*, 30 (1991), 351–76.

Paul, Herbert, 'The revived Channel Tunnel project', *The Nineteenth Century and After*, 61 (1907), 182–3.

Rapp, Dean, 'The early discovery of Freud by the British general educated public, 1912–1919', *Social History of Medicine*, 3 (1990), 217–43.

Rivers, W.H.R., 'Freud's psychology of the unconscious', *Lancet*, 1 (1917), 912–14.

Sanborn, Herbert, 'Why the Teuton fights', *The Fatherland. A Weekly, Germany's Just Cause*, New York, n.d., pp. 26–31.

Sartiaux, Albert, 'Le Tunnel sous-marin entre la France et l'Angleterre', *Revue des deux mondes*, 17 (1913), 543–75.

Schweik, Susan, 'Writing war poetry like a woman', *Critical Inquiry*, 13 (1987), 532–56.

Segal, Hanna, 'Silence is the real crime', *International Review of Psycho-Analysis*, 14 (1987), 3–12.

Steinberg, Jonathan, 'The Copenhagen complex', *Journal of Contemporary History*, 1 (1966), 23–46.

Stoddard, John L., 'Vigorous defence of Germany', *The Fatherland. A Weekly, Germany's Just Cause*, New York, n.d., pp. 38–42.

Summers, Anne, 'Militarism in Britain before the Great War', *History Workshop Journal*, 2 (1976), 104–23.

Sutherland, Alexander, 'The natural decline of warfare', *The Nineteenth Century*, 45 (1899), 570–8.

The Times, miscellaneous items, 1880–1920.

Valéry, Paul, 'La Conquête allemande', *The New Review*, 16 (1897), 99–112.

Watson, J.B., 'Psychology as the behaviourist views it', *Psychological Review*, 20 (1913), 158–77.

Willis, Kirk, 'The introduction and critical reception of Hegelian thought in Britain 1830–1900', *Victorian Studies*, 32 (1988), 85–111.

Wollen, Peter, 'Cinema/Americanism/the robot', *New Formations*, 8 (1989), 7–34.

Wright, D.G., 'The Great War, government propaganda and English "men of letters" 1914–16', *Literature and History*, 7 (Spring 1978), 70–100.

BOOKS

Alderson, E.A.H., *Lessons from 100 Notes Made in Peace and War*, Aldershot, 1908.

Anderson, Benedict, *Imagined Communities. Reflections on the Origin and Spread of Nationalism*, 1983.

Anon., *Essai sur la philosophie de la guerre. Evénements de 1870–1871*, Paris, 1872.

Aron, Raymond, *Clausewitz, Philosopher of War* [1976], translated by Christine Booker and Norman Stone, 1983.

Avineri, Shlomo, *Hegel's Theory of the Modern State*, Cambridge, 1972.

Ayling, R. Stephen, *Public Abattoirs: Their Planning, Design and Equipment*, 1908.

Baldick, Chris, *The Social Mission of English Criticism 1848–1932*, Oxford, 1983.

——— *In Frankenstein's Shadow. Myth, Monstrosity, and Nineteenth-Century Writing*, Oxford, 1987.

Barker, Ernest, *National Character and the Factors in its Formation*, 1927.

Barnett, Correlli and Slater, Humphrey, with the collaboration of R.H. Géneau, *The Channel Tunnel*, 1958.

Baroli, Marc, 'Le Train dans la littérature française', PhD thesis Paris University, 1963.

Barraclough, G., *From Agadir to Armageddon. Anatomy of a Crisis*, 1982.

Barrell, John, *The Infection of Thomas De Quincey: A Psychopathology of Imperialism*, New Haven and London, 1991.

Barthes, Roland, *Mythologies* [1957], translated by A. Lavers, 1973.

Bauman, Zygmunt, *Modernity and the Holocaust*, Cambridge, 1989.

Becker, Jean-Jacques, *1914: Comment les Français sont entrés dans la guerre. Contribution à l'étude de l'opinion publique printemps–été 1914*, Paris, 1977.

——— 'La Genèse de l'Union Sacrée', Colloquium, *1914. Les Psychoses de guerre*, Rouen, 1979, pp. 205–16.

Beer, Gillian, 'The island and the aeroplane', in Homi K. Bhabha (ed.), *Nation and Narration*, 1990, chapter 15.

Berg, Maxine, *The Machinery Question and the Making of Political Economy 1815–1858*, Cambridge, 1980.

Bernhardi, Friedrich von, *Germany and the Next War*, translated by Allen H. Powles, 1912.

——— *On War of To-day*, vol. I, *Principles and Elements of Modern War*, 1912.

——— *How Germany Makes War*, 1914.

——— *Notre Avenir*, Paris, 1915.

——— *The War of the Future in the Light of the Lessons of the World War*, translated by F.A. Holt, 2nd edn, 1920.

——— *La Guerra del Futuro*, Buenos Aires, 1921.

Best, Geoffrey, *Humanity in Warfare. The Modern History of the International Law of Armed Conflicts*, 1983.

Blackbourne, David, and Eley, Geoff, *The Peculiarities of German History. Bourgeois Society and Politics in Nineteenth-Century Germany*, Oxford, 1984.

Blanning, Tim, *The Origins of the French Revolutionary Wars*, London and New York, 1986.

Bloch, Jean de, *Impossibilités techniques et économiques d'une guerre entre les grandes puissances*, Paris, 1899.

Bonavia, Michael R., *The Channel Tunnel Story*, Newton Abbot, 1987.

Bonnal, H., *L'Esprit de la guerre moderne de Rosbach a Ulm*, Paris, 1903.
Borch-Jacobsen, Mikkel, *The Freudian Subject*, translated by Catherine Porter, 1989.
Bouthoul, Gaston, *Essai de polémologie. Guerre ou paix?*, Paris, 1976.
Bouthoul, Gaston, and Carrere, René, *Le Défi de la guerre (1740–1974). Deux siècles de guerres et de révolutions*, Paris, 1976.
Boutroux, E., *L'Idée de liberté en France et en Allemagne*, Paris, 1916.
Bracco, Rosa Maria, 'British middlebrow writers and the First World War, 1919–39', PhD, Cambridge, 1989.
Brewer, John, *The Sinews of Power: War, Money and the English State, 1688–1783*, 1990.
Briggs, Asa, 'The language of "mass" and "masses" in nineteenth-century England', *Ideology and the Labour Movement. Essays Presented to John Saville*, edited by David E. Martin and David Rubinstein, 1979, pp. 62–83.
Brock, Peter, *Pacifism in Europe to 1914*, Princeton, 1972.
—— *The Roots of War Resistance. Pacifism from the Early Church to Tolstoy*, New York, 1981.
Brook, Peter, *Reading for the Plot. Design and Intention in Narrative*, Oxford, 1984.
Brush, Stephen, *The Temperature of History. Phases of Science and Culture in the Nineteenth Century*, New York, 1978.
Butler, Samuel, *Erewhon and Erewhon Revisited* [1872 and 1901], Everyman edn, London 1932.
Čapek, Karel, *R.U.R.* [1920], New York, 1923.
—— *War with the Newts* [1936], New York, 1959.
Carlyle, Thomas, *Sartor Resartus*, in *Collected Works, 1885–1891*, vol. III.
Cash, C. and Heiss, H., *Our Slaughter House System. A Plea for Reform, and The German Abattoir*, 1907.
Challener, R., *The French Theory of the Nation in Arms 1866–1939*, New York, 1955.
Childers, Erskine, *The Riddle of the Sands* [1903], 1972.
Clarke, I.F., *Voices Prophesying War 1763–1984*, 1966.
Clausewitz, Carl von, *Principles of War* [1812], translated and edited with an introduction by Hans W. Gatzke, 1943.
—— *On War* [1832], edited and translated by Michael Howard and Peter Paret, Princeton, 1976.
—— *On War* [1832], edited with an introduction by Anatol Rapoport, Harmondsworth, 1986.
—— *Historical and Political Writings*, edited and translated by Peter Paret and Daniel Moran, Princeton, 1992.
Cobden, Richard, *Political Writings*, 2 vols, London and New York, 1867.
—— *Speeches on Questions of Public Policy*, edited by John Bright and James E. Thorold Rogers, 2 vols, 1870.
Cogniot, Georges, *Proudhon et la démagogie bonapartiste*, Paris, 1958.
Colley, Linda, 'Radical patriotism in eighteenth-century England', in *Patriotism: The Making and Unmaking of British National Identity*, edited by Raphael Samuel, 1989, 3 vols, vol. I, pp. 169–87.
Colomb, Philip H., Maurice, J.F., Maude, F.N., Forbes, A., Lowe, C., Murray, D.C. and Scudamore, F., *The Great War of 189–. A Forecast*, 1893.
Cottrell, Stella, 'The devil on two sticks: Francophobia in 1803', in *Patriotism: The Making and Unmaking of British National Identity*, edited by Raphael Samuel, 1989, 3 vols, vol. I, pp. 259–74.

Crile, George W., *A Mechanistic View of War and Peace*, n.d. [1916?].
Cruickshank, John, *Variations on Catastrophe. Some French Responses to the Great War*, Oxford, 1982.
Darwin, Charles, *The Origin of Species* [1859], Harmondsworth, 1987.
Deleuze, Gilles and Guattari, Felix, *Capitalisme et schizophrénie. Mille plateaux*, Paris, 1980.
——— *Nomadology: The War Machine*, New York, 1986.
De Potter, Agathon, *Qu'est-ce que la Guerre et la paix? Examen de l'ouvrage de M. Proudhon sur la guerre et la paix*, Brussels, 1862.
De Quincey, Thomas, *Suspiria De Profundis. Being a Sequel to 'The Confessions of an Opium Eater'*, in *Thomas De Quincey's Writings*, 20 vols, edited by J.T. Fields, Boston, 1851, vol. I.
——— 'The English Mail Coach; or the Glory of Motion', in *Thomas De Quincey's Writings*, vol. III.
——— 'On War', in *Thomas De Quincey's Writings*, Boston, 1854, vol. VIII.
Derrida, Jacques, *Positions* [1972], translated by Alan Bass, 1981.
De Riez, M.G. Martiny, *Histoire illustrée de la guerre de 1870–71, et de la guerre civile à Paris. Réflexions morales et politiques*, Laon (Aisne), 1871.
Digeon, Claude, *La Crise allemande de la pensée française (1870–1914)*, Paris, 1959.
Disraeli, Benjamin, *Sybil or the Two Nations* [1845], World's Classics, Oxford, 1981.
Douglas's Encyclopaedia. A Book of Reference for Bacon Curers . . . and all other industries associated with meat, pork, provision and general food trades, n.d. [1905?].
Driou, A., *Le Calvaire de la patrie. Invasion en France des hordes prussiennes*, Limoges, 1873.
Earle, Edward Mead (ed.), *Makers of Modern Strategy. Military Thought from Machiavelli to Hitler*, Princeton, 1944.
Eby, Cecil D., *The Road to Armageddon. The Martial Spirit in English Popular Literature 1870–1914*, Durham and London, 1988.
Eder, M.D., *War Shock. The Psycho-neuroses in War Psychology and Treatment*, 1917.
Eisenstein, S.M., *The Complete Films of Eisenstein together with an Unpublished Essay by Eisenstein*, translated by John Hetherington, n.d.
Eksteins, Modris, *Rites of Spring. The Great War and the Birth of the Modern Age*, 1989.
Emerson, E.W. (ed.), *A Correspondence between John Sterling and Ralph Waldo Emerson*, Boston, 1897.
Encyclopaedia Britannica, 'War', 11th edn, 1910–11, vol. XXVIII, 305–16.
Engels, Friedrich, *Anti-Dühring. Herr Eugen Dühring's Revolution in Science* [1878], translated by Emile Burns, Moscow, 1947.
——— *Notes on the War. Sixty Articles Reprinted from the Pall Mall Gazette 1870–1871*, translated by Friedrich Adler, Vienna, 1923.
Eysenck, Hans, *The Decline and Fall of the Freudian Empire*, Harmondsworth, 1985.
Ferenczi, S., Abraham, K., Simmel, E. and Jones, E., *Psychoanalysis and the War Neuroses*, introduced by Sigmund Freud, London, Vienna and New York, 1921.
Ferro, Marc, *The Great War 1914–1918* [1969], 1987.
Fischer, Fritz, *World Power or Decline. The Controversy over Germany's Aims in the First World War* [1965], translated by L.L. Farrar, Robert Kimber and Rita Kimber, 1975.

Florence, Mary Sargant, Marshall, Catherine and Ogden, C.K, *Militarism versus Feminism. Writings on Women and War* [1915], 1987.

Fornari, Franco, *The Psychoanalysis of War* [1966], translated by Alenka Pfeifer, Bloomington, 1975.

Forster, E.M. 'The Machine Stops', in *Collected Short Stories*, 1948.

Foucault, Michel, *Folie et déraison. Histoire de la folie*, Paris, 1961.

—— *Discipline and Punish. The Birth of the Prison* [1975], translated by Alan Sheridan, Harmondsworth, 1977.

—— *Power/Knowledge: Selected Interviews*, edited by Colin Gordon, Brighton and New York, 1980.

—— *Foucault Live*, New York, 1989.

Freud, Sigmund, *The Psychopathology of Everyday Life* [1901], in *The Standard Edition of the Complete Psychological Works of Sigmund Freud* [SE] translated from the German under the General Editorship of James Strachey in collaboration with Anna Freud, assisted by Alix Stratchey and Alan Tyson.

—— 'The future prospects of psycho-analytic therapy', [1910], *SE*, vol. XI.

—— 'On the history of the psycho-analytic movement' [1914], *SE*, vol. XIV.

—— 'Thoughts for the times on war and death' [1915], *SE*, vol. XIV.

—— 'On transience', [1916], *SE*, vol. XIV.

—— *Introductory Lectures on Psycho-Analysis* [1916–17], *SE*, vols XV–XVI.

—— 'Introduction to *Psycho-Analysis and the War Neuroses*' [1919], *SE*, vol. XVII.

—— 'Memorandum on the electrical treatment of war neurotics' [1920, published 1955], *SE*, vol. XVII.

—— *Beyond the Pleasure Principle* [1920], *SE*, vol. XVIII.

—— *Group Psychology and the Analysis of the Ego* [1921], vol. XVIII.

—— 'Psycho-analysis and telepathy' [1921, published 1941], *SE*, vol. XVIII.

—— 'An autobiographical study' [1925], *SE*, vol. XX.

—— *Civilisation and its Discontents* [1930], *SE*, vol. XXI.

—— *Why War?* [1933], *SE*, vol. XXII.

—— *Analysis Terminable and Interminable* [1937], *SE*, vol. XXIII.

—— *Moses and Monotheism* [1939], *SE*, vol. XXIII.

—— *A Phylogenetic Fantasy. Overview of the Transference Neuroses*, edited with an essay by Ilse Grubrich-Simitis, translated by Axel Hoffer and Peter T. Hoffer, Cambridge, Mass., 1987.

Freud, Sigmund and Abraham, Karl, *A Psycho-Analytic Dialogue. The Letters of Sigmund Freud and Karl Abraham 1907–1926*, edited by Hilda C. Abraham and Ernst L. Freud, translated by Bernard Marsh and Hilda C. Abraham, New York, 1965.

Freud, Sigmund and Andreas-Salomé, Lou, *Letters*, edited by Ernst Pfeiffer, translated by W. and E. Robson-Scott, 1972.

Fromm, Erich, *The Fear of Freedom*, 1942.

—— *The Anatomy of Human Destructiveness*, New York, Chicago and San Francisco, 1973.

Fuller, J.F.C., *The Foundations of the Science of War*, 1926.

—— *War and Western Civilization 1832–1932. A Study of War as a Political Instrument and the Expression of Mass Democracy*, 1932.

Fussell, Paul, *The Great War and Modern Memory*, Oxford, 1975.

Gallie, W.B., *Philosophers of War and Peace*, Cambridge, 1978.

—— *Understanding War*, 1991.

Gat, Azar, *The Origins of Modern Military Thought. From the Enlightenment to Clausewitz*, Oxford, 1989.

Gay, Peter, *Freud: A Life for Our Time*, 1988.
Geyl, Peter, *Napoleon: For and Against* [1949], Harmondsworth, 1986.
Giddens, Anthony, *The Nation State and Violence*, Cambridge, 1985.
Giedion, Siegfried, *Mechanization takes Command. A Contribution to Anonymous History*, New York, 1948.
Gilman, Sander, *The Jew's Body*, 1991.
Gleason, John Howes, *The Genesis of Russophobia in Great Britain*, Cambridge, Mass., 1950.
Glover, Edward, *War, Sadism and Pacifism* [1933], enlarged edn, 1947.
Glucksmann, André, *Le Discours de la guerre* [1967], Paris, 1974.
Gobineau, Arthur Comte de, *Selected Political Writings*, edited and introduced by Michael Biddis, 1970.
Hagerman, Edward, *The American Civil War and the Origins of Modern Warfare. Ideas, Organization, and Field Command*, Bloomington and Indianapolis, 1988.
Halévy, Élie, *The Era of Tyrannies* [1938], translated by R.K. Webb, 1967.
Hamerow, Theodore S., *The Social Foundations of German Unification 1858–1971. Ideas and Institutions*, Princeton, 1969.
Hearnshaw, F.J.C., *Germany the Aggressor throughout the Ages*, Foreword by Sir Thomas H. Holland, London and Edinburgh, 1940.
Hegel, G.W.F., *The Phenomenology of Spirit* [1807], translated by A.V. Miller, Oxford, 1979.
—— *The Philosophy of Right* [1821], translated by T.M. Knox, Oxford University Press, 1967.
—— *Lectures on the Philosophy of World History* [1822–30], translated by H.B. Nisbet, Cambridge, 1975.
Henderson, G.F.R., *The Science of War. A Collection of Essays and Lectures 1892–1903*, edited by N. Malcolm, 1905.
Hennebert, Eugène, *L'Art militaire et la science*, Paris, 1884.
—— *La Guerre imminente. La Défense du territoire*, Paris, n.d. [1890?].
Hibbert, Christopher, *The Destruction of Lord Raglan. A Tragedy of the Crimean War 1854–1855* [1961], Harmondsworth, 1984.
Higonnet, M.R., Jenson, J., Michel, S. and Weitz, M.C., *Behind the Lines. Gender and the Two World Wars*, New Haven and London, 1987.
Hichberger, J., 'Old soldiers', in *Patriotism: The Making and Unmaking of British National Identity*, edited by Raphael Samuel, 1989, 3 vols, vol. III, pp. 50–64.
Hobhouse, L.T., *Questions of War and Peace*, 1916.
Hobsbawm, Eric, *Industry and Empire*, Harmondsworth, 1969.
—— *Nations and Nationalism since 1780. Programme, Myth, Reality*, Cambridge, 1990.
Hobson, J.A., *The Psychology of Jingoism*, 1901.
—— *The Evaluation of Modern Capitalism. A Study of Machine Production* [1894], revised edn, 1926.
Hodgen, Margaret, *The Doctrine of Survivals. A Chapter in the History of Scientific Method in the Study of Man*, 1936.
Hoffman, Robert L., *Revolutionary Justice. The Social and Political Theory of P.-J. Proudhon*, Urbana, 1972.
Howard, Michael, *Studies in War and Peace* [1959], 1970.
—— *The Franco-Prussian War. The German Invasion of France, 1870–1871* [1961], 1988.
—— 'The influence of Clausewitz', introductory essay in Clausewitz's *On War*, edited and translated by Howard and Paret, Princeton, 1976.
—— *The Causes of War and Other Essays*, Hounslow, 1983.

Hynes, Samuel, *A War Imagined. The First World War and English Culture*, 1990.
Jacomet, Robert, *Les Lois de la guerre continentale*, Paris, 3rd edn, 1913.
James, Henry, *The Letters of Henry James*, selected and edited by Percy Lubbock, 2 vols, 1920.
Joad, C.E.M., *Why War?*, Harmondsworth, 1939.
Jomini, Antoine-Henri de, *Précis de l'art de la guerre*, St Petersburg, 1837.
Jones, Ernest, *Sigmund Freud: Life and Work*, 3 vols, 1955.
Kant, Immanuel, *Political Writings*, edited with an introduction by Hans Reiss, Cambridge, 1970.
Keegan, John, *The Face of Battle*, Harmondsworth, 1986.
Kennedy, Paul M., *The Rise of the Anglo-German Antagonism 1860–1914*, 1982.
Kern, Stephen, *The Culture of Time and Space 1880–1918*, 1983.
Kevles, Daniel, *In the Name of Eugenics. Genetics and the Uses of Human Heredity*, Harmondworth, 1985.
Kipling, Rudyard, *The War and A Fleet in Being*, in *The Sussex Edition of the Complete Works in Prose and Verse of Rudyard Kipling*, vol. XXVI, 1938.
Klein, Melanie and Riviere, Joan, *Love, Hate and Reparation* [1937], 1967.
Koch, H.W., 'Social Darwinism as a factor in the "New Imperialism"', in Koch (ed.), *The Origins of the First World War. Great Power Rivalry and German War Aims*, 1972, pp. 329–54.
Koestler, Arthur, *The Ghost in the Machine*, 1967.
Kojève, Alexandre, *Introduction to the Reading of Hegel. Lectures on the Phenomenology of Spirit*, assembled by Raymond Quesneau, edited by Allan Bloom, translated by James H. Nichols, Jr, London and New York, 1969.
Koonz, Claudia, *Mothers in the Fatherland. Women, the Family and Nazi Politics*, 1986.
Kristeva, Julia, *Étrangers à nous-mêmes*, Paris, 1988.
Larousse, *Grand Dictionnaire universel du XIXe siècle*, Paris, 1872.
Larrieu, A., *Guerre à la guerre*, 3rd edn, Paris, 1868.
Larroque, M.P., *De la Guerre et des armées permanentes*, Paris, 1856.
Latour, Bruno, *The Pasteurization of France* [1984], translated by Alan Sheridan and John Law, Cambridge, Mass., 1988.
Laurent-Chirlonchon, M.V., *Des Vraies Causes de la supériorité de la Prusse en 1866 et en 1870*, Paris, 1875.
La Vergata, Antonello, *L'equilibrio e la guerra della natura. Dalla teologia naturale al darwinismo*, Naples, 1990.
Lavisse, E. and Andler, C., *German Theory and Practice of War*, Paris, 1915.
Le Bon, Gustave, *The Psychology of the Great War*, 1916.
Lee, Gerald Stanley, *The Voice of the Machines. An Introduction to the Twentieth Century*, New York, 1906.
Leed, Eric, *No Man's Land. Combat and Identity in World War I*, Cambridge, 1979.
Lenin, V.I., 'Position and tasks of the Socialist International', *Collected Works*, vol. XXI [August 1914–December 1915], Moscow, 1964.
—— *Socialism and War* [1915], in *Collected Works*, vol. XXI.
—— *Imperialism, the Highest Stage of Capitalism* [1917], Moscow, 1982.
Le Queux, William, *The Great War in England in 1897*, 1894.
—— *The Invasion of 1910 with a Full Account of the Siege of London*, 1906.
Letourneau, Charles-Jean, *La Guerre dans les diverses races humaines*, Bibliothèque anthropologique, Paris, 1895.
Luxemburg, Rosa, *The Letters of Rosa Luxemburg*, edited and introduced by Stephen Eric Bronner, Boulder, Colorado, 1978.

—————— 'The crisis in German Social Democracy (*The Junius Pamphlet: Part One*)', in *Selected Political Writings of Rosa Luxemburg*, edited and introduced by Dick Howard, New York and London, 1971.

McCabe, Joseph, *Treitschke and the Great War*, 1914.

McConville, Sean, *A History of English Prison Administration*, vol. I, *1750–1877*, 1981.

Mackenzie, John M., *Propaganda and Empire. The Manipulation of British Public Opinion 1880–1960*, Manchester, 1984.

Marinetti, F.T., 'Sintesi futurista della guerra', in *Teoria e invenzione futurista, Opere*, Milan, 1968, vol. II, pp. 280–1.

Marx, Karl, *The Communist Manifesto* [1848], edited by Frederick L. Bender, New York, 1988.

—————— *Grundrisse. Foundation of the Critique of Political Economy (Rough Draft)*, translated by M. Nicolaus, Harmondsworth, 1973.

Marx, Karl and Engels, Friedrich, *Correspondence. A Selection, 1846–1895*, 1934.

Mayhew, Henry, *London Labour and the London Poor. A Cyclopaedia of the Condition and Earnings of Those That Will Work, Those That Cannot Work, and Those That Will Not Work* [1861–2], 4 vols, 1967.

Middlebrook, Martin, *The First Day on the Somme. 1 July 1916*, Harmondsworth, 1984.

Miller, J. Hillis, *The Disappearance of God. Five Nineteenth-Century Writers* [1963], Cambridge, Mass., 1975.

Mommsen, Wolfgang, J., *Max Weber and German Politics 1890–1926* [1959], translated by Michael S. Steinberg, Chicago, 1984.

—————— 'The topos of inevitable war', in *Germany in the Age of Total War*, edited by Volker R. Berghahn and Martin Kettle, 1981, pp. 23–45.

Money-Kyrle, Roger, *Superstition and Society*, 1939.

—————— *Collected Papers*, edited by Donald Meltzer with the assistance of Mrs Edna O'Shaughnessy, Strath Tay, 1978.

Monod, Gabriel, *Allemands et Français*, Paris, 1872.

Mulhern, Francis, *The Moment of 'Scrutiny'*, 1979.

Nicolai, G.F., *The Biology of War* [*Die Biologie des Krieges*, Zurich, 1917] translated by Constance and Julian Grande, 1919.

Nora, Pierre (ed.), *Les Lieux de mémoire*, Paris, 1984.

Nye, Robert, *The Origins of Crowd Psychology: Gustave Le Bon and the Crisis of Mass Democracy in the Third Republic*, London and Beverley Hills, 1975.

Paret, Peter, 'The genesis of *On War*', introductory essay in Clausewitz's *On War*, edited and translated by Howard and Paret, Princeton, 1976.

Pearson, Karl, *National Life from the Standpoint of Science* [1901], 2nd edn, 1905.

—————— *The Scope and Importance to the State of the Science of National Eugenics*, 3rd edn, 1911.

Peel, J.D.Y., *Herbert Spencer. The Evolution of a Sociologist*, 1971.

Pemberton, Max, *Pro Patria*, illustrated by A. Forrestier, 1901.

Playne, Caroline, E., *The Neuroses of the Nations*, 1925.

—————— *The Pre-War Mind in Britain. An Historical Review*, 1928.

Pocock, J.G.A., *The Machiavellian Moment. Florentine Political Thought and the Atlantic Republican Tradition*, Princeton, 1975.

Poliakov, Léon, *The Aryan Myth. A History of Racist and Nationalist Ideas in Europe*, translated by E. Howard, 1974.

Poster, Mark, *Existential Marxism in Post War France: From Sartre to Althusser*, Princeton, 1975.

Proudhon, P.-J., *La Guerre et la paix. Recherches sur le principe et la constitution du droit des gens* [1861], in *Oeuvres complètes*, edited by C Bougle and H. Moysset, vol. VI, Paris, 1927.

────── *Carnets de P.-J. Proudhon*, edited by Pierre Haubtmann, vol. III (1848–50), Paris, 1968.

Quatrefages, Jean Louis Armand de, *The Prussian Race Ethnologically Considered to which is appended some account of the bombardment of the Museum of Natural History etc., by the Prussians in January 1871*, translated by Isabella Innes, 1872.

Quiller-Couch, Arthur, *On the Art of Reading. Lectures Delivered in the University of Cambridge 1916–1917*, Cambridge, 1920.

Raleigh, Walter, *England and the War. Being Sundry Addresses Delivered during the War*, Oxford, 1918.

Read, James Morgan, *Atrocity Propaganda 1914–1919*, New Haven, 1941.

Reich, Wilhelm, *The Mass Psychology of Fascism* [1933], 3rd edn, New York, 1946.

Remarque, Erich Maria, *All Quiet on the Western Front*, translated by A.W. Wheen, 1929.

Renan, Ernest, 'What is a nation?' [1882], in Homi K. Bhabha (ed.), *Nation and Narration*, 1990, chapter 2.

Richet, Charles, *Le Passé de la guerre et l'avenir de la paix*, Paris, 1907.

Richter, Melvin, *The Politics of Conscience. T.H. Green and His Age*, 1964.

Rilke, Rainer Maria, *Briefe aus den Jahren 1914 bis 1921*, edited by Ruth Sieber-Rilke and Carl Sieber, Leipzig, 1938.

────── *War Time Letters*, translated by M.D. Herter Norton, New York, 1940.

Ritchie, D.G., *Darwin and Hegel, with Other Philosophical Studies*, 1893.

Ritvo, Lucille B., *Darwin's Influence on Freud. A Tale of Two Sciences*, New Haven, 1990.

Roazen, Paul, *Freud and his Followers*, Harmondsworth, 1979.

Rolland, Romain, and Masereel, Frans, *La Révolte des machines ou la pensée dechainée* [1921], Paris, 1947.

Ronsin, Francis, *La Grève des ventres. Propagande néo-malthusienne et baisse de la natalité française*, Paris, 1980.

Rose, Jacqueline, *The Haunting of Sylvia Plath*, 1991.

────── *Why War? Psychoanalysis, Politics and the Return to Melanie Klein*, Bucknell Lectures in Literary Theory, Oxford, 1993.

Rosenberg, John D., *The Darkening Glass. A Portrait of Ruskin's Genius*, New York, 1986.

Roth, Michael S., *Knowing and History. Appropriations of Hegel in Twentieth-Century France*, Ithaca, 1988.

Rothenberg, Gunther E., *The Art of Warfare in the Age of Napoleon*, 1977.

Roudinesco, Elizabeth, *La Bataille de cent ans. Histoire de la psychanalyse en France*, vol. I, *1885–1939*, vol. II, *1925–1985*, Paris 1986. Second volume translated by Jeffrey Mehlman as *Jacques Lacan and Co. A History of Psychoanalysis in France, 1925–1985*, 1990.

Ruskin, John, *The Crown of Wild Olive* [1866], in *The Works of John Ruskin*, edited by E.T. Cook and Alexander Wedderburn, 1903–12, vol. XVIII.

Sesame and Lilies [1865, 1871], in *The Works of John Ruskin*, vol. XVIII.

────── *Fors Clavigera. Letters to the Workmen and Labourers of Great Britain*

[1871–84], in *The Works of John Ruskin*, vol. XXVII.
—— *Munera Pulveris. Six Essays on the Elements of Political Economy* [1872], in *The Works of John Ruskin*, edited by E.T. Cook and Alexander Wedderburn, 1903–12, vol. XVII.
—— *The Storm Cloud of the Nineteenth Century. Two Lectures Delivered at the London Institution*, Orpington, 1884.
Russell, Bertrand, *The Autobiography of Bertrand Russell*, 1968.
Saki [H.H. Munro], *When William Came. A Story of London under the Hohenzollerns* [1913], in *The Penguin Complete Saki*, Harmondsworth, 1982.
Samuel, Raphael (ed.), *Patriotism: The Making and Unmaking of British National Identity*, 3 vols, 1989.
Sanders, M.L. and Taylor, Philip M., *British Propaganda during the First World War, 1914–18*, 1982.
Schama, Simon, *Citizens. A Chronicle of the French Revolution*, Harmondsworth, 1989.
Schulz, Albert, *Bibliographie de la guerre franco-allemande (1870–1871) et de la Commune de 1871*, Paris, 1886.
Schumpeter, Joseph A., *Imperialism and Social Classes* [1919 and 1927], Oxford, 1951.
—— *Capitalism, Socialism and Democracy*, 1943.
Schwarz, Oscar, *Public Abattoirs and Cattle Markets*, 2nd edn, 1902.
Semmel, Bernard, *Marxism and the Science of War*, 1981.
Seton Watson, R.W., Dover Wilson, J., Zimmern, Alfred E. and Greenwood, Arthur, *The War and Democracy*, 1914.
Shelley, Mary, *Frankenstein, or the Modern Prometheus* [1818], 1968.
Silberner, Edmund, *The Problem of War in Nineteenth Century Economic Thought*, translated by A.H. Krappe, Princeton, 1946.
Sinclair, Upton, *The Jungle* [1906], Harmondsworth, 1985.
Smith, A.L., 'The people and the duties of empire', in *The Empire and the Future*, edited by A.P. Newton, Imperial Studies Committee, University of London, 1916.
Spencer, Herbert, *The Principles of Sociology*, 3 vols, 1876–96.
Stepan, Nancy Leys, ' "Nature's pruning hook": war, race and evolution, 1914–18', in J.M.W. Bean (ed.), *The Political Culture of Modern Britain: Studies in Memory of Stephen Koss*, 1987, pp. 129–48.
Stepansky, Paul E., *A History of Aggression in Freud*, New York, 1977.
Steward, Bertrand, *The Active Service Pocket Book* [1906], 5th edn, 1912.
Stoddard, Lothrop, *The Rising Tide of Color against White World-Supremacy*, 1920.
Stone, M. 'Shellshock and the psychologists', in *The Anatomy of Madness*, edited by W.F. Bynum, R. Porter and M. Shepherd, 1985, vol. II, 242–71.
Strachey, Alix, *The Unconscious Motives of War*, 1957.
Stromberg, Roland N., *Redemption by War. The Intellectuals and 1914*, Kansas, 1982.
Sulloway, Frank, *Freud, Biologist of the Mind. Beyond the Psychoanalytic Legend*, New York, 1979.
Surel, Jeanine, 'John Bull', in *Patriotism: The Making and Unmaking of British National Identity*, edited by Raphael Samuel, 1989, 3 vols, vol. III, pp. 3–25.
Tarde, Gabriel, *Underground Man* [1896], preface by H.G. Wells, 1905.
Taylor, Frederick Winslow, *The Principles of Scientific Management*, New York, 1911.
Teich, Mikuláš and Porter, Roy (eds), *Fin-de-siècle and its Legacy*, Cambridge, 1990.

Terraine, John, *White Heat. The New Warfare 1914–18*, 1982.

Theweleit, Klaus, *Male Fantasies*, 2 vols, Cambridge, 1987 and 1989.

Thom, Martin, 'Tribes within nations: the ancient Germans and the history of modern France', in Homi K. Bhabha (ed.), *Nation and Narration*, 1990, chapter 3.

Timms, Edward, *Karl Kraus: Apocalyptic Satirist. Culture and Catastrophe in Habsburg Vienna*, New Haven, 1986.

Tolstoy, Leo, *Tales of Sevastopol. The Cossacks*, Moscow, n.d. [1983?].

—— *War and Peace* [1863–9], translated by Louise and Aylmer Maude, World's Classics, Oxford, 1954.

Treitschke, Heinrich von, *L'Avenir des moyens états du Nord de l'Allemagne*, Paris, 1866.

—— *The Fire-Test of the North-German Confederation*, translated by Frederick A. Hyndman, 1870.

—— *The Confessions of Frederick the Great and the Life of Frederick the Great*, 1914; New York, 1915.

—— *The Organisation of the Army. Being 23 of his Lectures on Politics*, translated by Adam A. Gowans, 1914.

—— *Germany, France, Russia and Islam*, 1915.

—— *History of Germany in the Nineteenth Century*, translated by Eden and Cedar Paul, introduced by W.H. Dawson, 7 vols, 1915–1919.

—— *Politics*, translated by Blanche Dugdale and Torben De Bille, with an introduction by Arthur James Balfour, 2 vols, 1916.

—— *La Francia dal Primo Impero al 1871*, translated by Enrico Ruta, Bari, 1917.

—— *Treitschke's Origins of Prussianism (The Teutonic Knights)*, translated by Eden and Cedar Paul, 1942.

Valéry, Paul, *Une Conquête méthodique* [1897], Abbeville, 1924.

Veblen, Thorstein, *Imperial Germany and the Industrial Revolution*, New York, 1915.

Veith, Ilza, *Hysteria: the History of a Disease*, Chicago, 1965.

Verene, D.P., 'Hegel's account of war', in Z.A. Pelczynski (ed.), *Hegel's Political Philosophy. A Collection of New Essays*, Cambridge, 1971, pp. 168–80.

Vialles, Noélie, *Le Sang et la chair. Les abattoirs des pays de l'Adour*, Paris, 1987.

Virilio, Paul, *War and Cinema. The Logistics of Perception*, translated by Patrick Camiller, 1989.

Vondung, Klaus, 'Visions de mort et de fin de monde: attente et désir de la guerre dans la littérature allemande avant 1914', in *Colloquium, 1914. Les Psychoses de guerre*, Rouen, 1979, pp. 217–45.

Wallace, Stuart, *War and the Image of Germany. British Academics 1914–1918*, Edinburgh, 1988.

Watson, J.B., *Behaviour. An Introduction to Comparative Psychology*, New York, 1914.

—— *Behaviourism*, 1928.

Weart, Spencer R., *Nuclear Fear. A History of Images*, Cambridge, Mass., 1988.

White, Arnold, *Problems of a Great City*, 1886.

Willey, Basil, *Cambridge and Other Memories 1920–1953*, 1968.

Williams, Gordon, '"Remember the Llandovery Castle": cases of atrocity propaganda in the First World War', in *Propaganda, Persuasion and Polemic*, edited by Jeremy Hawthorn, 1987, pp. 19–34.

Williams, Rosalind, *Notes on the Underground. An Essay on Technology, Society, and the Imagination*, Cambridge, Mass., 1990.

Woodcock, George, *Pierre-Joseph Proudhon. A Biography*, 1956.
Woolf, Virginia, *Three Guineas* [1938], 1991.
Zola, Emile, *La Bête humaine* [1890], translated by Leonard Tancock, Harmondsworth, 1977.

Index

War
Machine

Steve [signature]

Science Studies Unit
University of Edinburgh
Jan. 1996